Fatigue of Textile and Short Fiber Reinforced Composites

Durability and Ageing of Organic Composite Materials Set
coordinated by
Marco Gigliotti

Volume 1

Fatigue of Textile and Short Fiber Reinforced Composites

Valter Carvelli
Atul Jain
Stepan Lomov

iSTE

WILEY

First published 2017 in Great Britain and the United States by ISTE Ltd and John Wiley & Sons, Inc.

ISTE Ltd
27-37 St George's Road
London SW19 4EU
UK

www.iste.co.uk

John Wiley & Sons, Inc.
111 River Street
Hoboken, NJ 07030
USA

www.wiley.com

Library of Congress Control Number: 2017930848

British Library Cataloguing-in-Publication Data
A CIP record for this book is available from the British Library
ISBN 978-1-78630-021-8

Contents

Preface

Safety, durability and reliability considerations of composite material components necessitate an in-depth knowledge of different aspects of the materials behavior. In particular, composite components exposed to cyclic loading throughout their life-time could and usually do suffer from mechanical performance degradation, namely 'fatigue'. In the last three or four decades, intense and continuous investigation has been primarily devoted to the fatigue behavior of unidirectional laminated composites, as manufacturing of these long fiber reinforced plastics was suitable for the earlier applications in the aeronautic and naval industry. Nowadays, development of new resins, improvement of the impregnation techniques, and availability of a broad range of materials and architecture for reinforcements allow the application of composites for mass scale production of lightweight components. Therefore, there is high demand from various industries, such as the automotive, aeronautical, marine, and energy production industries, among others, for knowledge of the fatigue mechanisms and design of suitable fatigue resistant structural, as well as non-structural, composite elements subjected to long-term fluctuating loads.

In the broad family of fiber reinforced composite materials, textile reinforced and injection molded short fiber reinforced composites, in the Authors' opinion, have some mechanical and manufacturing peculiarities suitable for several applications, also covering long-term variable loadings.

The automated, computer controlled machines for industrial manufacturing of textile reinforcements and the increasing machine operation speed makes such preforms competitive from the productivity and affordability viewpoints. A huge variety of interlacement architectures for composite reinforcements, ranging from two- to three-dimensional fabrics, are now available. These possess excellent drapability and versatility, which is extremely important for complex double-curvature shape components.

Ease of manufacturing, low cost and reasonable specific mechanical properties make short fiber reinforced composites (SFRC) an attractive material for a large number of industrial applications, particularly in the automotive industry. The injection molding manufacturing process allows rapid and inexpensive fabrication of composite components with complex features like notches, curved surfaces and irregular thickness. This attribute, combined with reasonable specific properties, makes them very attractive for a wide range of semi-structural applications.

This book presents an overview of the fatigue behavior of these two different reinforced composites, namely textile composites (Part 1) and short fiber reinforced composites (Part 2).

Part 1, dedicated to textile composites, gathers several understandings on the fatigue imparted damage evolution. Chapter 1 presents experimental investigations of the mechanical behavior and the fatigue damage development in 2D and 3D textile reinforced composites. For each material, a complete description of the damage evolution is detailed covering the prefatigue quasi-static, the fatigue and the postfatigue quasi-static tensile behavior. The prefatigue quasi-static tensile loading provides the relevant mechanical properties and understanding of the initiation and development of the damage mechanisms. The tensile-tensile fatigue loading, accompanied by the fatigue life diagram, gives in-depth understanding on the damage modes sequence related to the reinforcement architecture. Moreover, analogies of the static and cyclic damage mechanisms are highlighted to obtain an initial connection of the damage thresholds and the “fatigue limit”, extensively discussed in Chapter 2. The postfatigue quasi-static behavior points out the effect of the fatigue on the mechanical properties and on the modification of the initiation and development of the damage modes.

The aim of Chapter 2 is to extend the notion of "fatigue limit" (namely, the cyclic load level below which the fatigue failure at a given large number of cycles does not happen) to textile composites. In particular, the relationship, if it exists, of the damage thresholds and the fatigue limit. The considered damage thresholds (i.e. the load levels distinguishing the different stages of the damage evolution) were identified, as in Chapter 1, using acoustic emission (AE) registration during the quasi-static tensile loading. The extensive dataset of various glass and carbon fiber reinforced thermoset textile composites confirms the link between the quasi-static damage thresholds and the fatigue life limit. However, the complex fatigue behavior of textile composites does not indicate a unique correlation of the fatigue limit and the quasi-static damage thresholds. This correlation is shown to be very different for glass and carbon reinforced textile composites.

Part 2 collects different perspectives on the fatigue response of short fiber reinforced composites. Chapter 3 compares a wide set of experimental measurements to highlight the dependency of the fatigue behavior on a large number of factors, including: fiber and matrix materials, fiber distribution, environmental and loading factors. Fatigue behavior of composites manifests itself in two easily observable trends. First is the SN curve; it represents the number of cycles to failure if a certain load is applied. Second is the loss of stiffness; SFRC materials suffer from continuous loss of stiffness during cyclic loading. Both these aspects are described in detail with emphasis on statistical comparisons to underline dependence on various factors. This chapter reinforces the different challenges one must overcome during the fatigue simulation of SFRC, some of which are dealt with in Chapter 4.

In Chapter 4, a component level simulation method for SFRC is detailed by introducing a new multiscale hybrid modelling concept for fatigue simulation. This proposed approach combines test results and simulation results (hybrid approach) on different scales (microscopic simulation, macroscopic fatigue behavior), overcoming the problems of depending on too many expensive tests (pure test-based macroscopic approach) and a lack of efficiency (pure microscopic simulation approach). Four different steps are identified for successful component level simulation. First, the appropriate homogenization scheme is recognized. Next, the micro-mechanics damage modelling of SFRC is built within the framework of the chosen homogenization scheme. The damage at the micro-scale is then related to the macro-scale fatigue properties by means of a Master SN curve approach. Finally, all of the elements are integrated with manufacturing simulation software and fatigue solvers to achieve component level simulation. Each of the four steps are described in detail with validation of the models and examination of the assumptions at every step of model development.

Fatigue of fiber reinforced composites, like a number of other engineering problems, offer two contrasting approaches. The first is that of a material scientist who would like to understand reality with the highest possible accuracy, the second is that of an engineer who desires to develop tools to safely design parts which might be subject to complex conditions. Though not mutually exclusive, every investigation of fatigue of composite materials has in general a certain bias towards one of the two approaches. In this book, we have tried to give a flavor of both approaches: Chapter 1 can be viewed as a scientists' approach towards the understanding of fatigue of textile composites; whereas Chapter 4 presents a predominantly engineers' approach to fatigue of composite materials. In both cases, however, the perspective of the other (scientist or engineer) has also been respected and dwelt upon at different points.

The results and understandings collected in the book are achievements of several years of fruitful collaborations with many colleagues and friends. We particularly acknowledge: Giulia Gramellini, Juan Pazmino and Vanni Neri Tomaselli (former masters students at Politecnico di Milano and Erasmus Visiting Scholars in KU Leuven); Yasmine Abdin, Jose Manuel Beas, Katleen Vallons and Ignaas Verpoest (KU Leuven); Alexander E. Bogdanovich and Dmitri D. Mungalov (former members of R&D department at 3Tex Inc.).

PART 1

Fatigue of Textile Composites

1

Fatigue Behavior and Damage Evolution of 2D and 3D Textile-Reinforced Composites

1.1. Introduction

The application of textile-reinforced composites in several industrial applications requires in-depth knowledge of the material's fatigue performance. Several parameters affect the fatigue response of textile composites: fiber and matrix material, fiber volume fraction, environmental conditions (temperature, moisture absorption, etc.), loading conditions (average stress, minimum to maximum stress ratio *R*, cycling frequency, etc.) and, last but not least, the architecture of the reinforcement. The latter and the material of fibers are mainly considered in this part of the book, acknowledging that the other factors are equally important.

Tension–tension loading is the most commonly used experimental test to investigate the influence of those parameters on the fatigue behavior of composite materials and, in particular, of textile-reinforced composites. Thus, only the tensile fatigue loading condition is considered in this chapter. We are aware that it is not an exhaustive, complete investigation of the fatigue response. Other cyclic loadings are quite important for several applications. Some of the loading conditions most investigated and available in the literature (see database collections [VAL 15]), are uniaxial tension–compression

[KAW 15], compression–compression [MOU 07] and multiaxial tension–torsion [INO 00, QUA 14].

As observed for some composite materials (see, e.g., [TAL 12, VAL 07]), the damage modes under tensile fatigue loading are analogous to those imparted with a quasi-static tensile load. This means that the effect of a quasi-static increment of the tensile force produces the same damage evolution as an increase in the number of fatigue cycles.

Accordingly, a complete fatigue investigation must include, in the authors' opinion, three steps. The first is an in-depth investigation of the prefatigue quasi-static tensile behavior for the knowledge of the relevant mechanical properties, and understanding of the initiation and development of the damage mechanisms. The second is a wide range, in term of load levels, of tensile cyclic loading for drawing a fatigue life diagram covering from low to high number of fatigue life cycles. Moreover, observation of the damage development during cyclic loading is supposed for comparison with the monitoring during quasi-static loading. Finally, the third step is dedicated to postfatigue quasi-static tensile tests of specimens previously subjected to different number of cycles and fatigue loading levels. This provides an overview on the effect of the fatigue loading on the degradation of the mechanical properties and on the modification of the initiation and development of the damage mechanisms.

The chapter gives an overview on the three steps investigation for two-dimensional (2D) and three-dimensional (3D) glass and carbon textile reinforced composites. In particular, the experimental results refer to epoxy resin reinforced with:

– plain weave E-glass textile (PWG);

– single-ply non-crimp 3D orthogonal weave E-glass textile (3DW);

– 3D rotary braided carbon textile (3DB);

– non-crimp stitched and unstitched carbon fabrics (NCFs).

This set of reinforcements covers a broad range. The general phenomena and features, identified in the studies reported in the chapter, are likely to be present in other instances of textile-reinforced composites, encountered in various applications (aeronautic, automotive, wind energy, etc.).

For each composite, after the features of the material and some peculiar experimental details, a comprehensive description is presented dealing with the three steps described namely:

– the main quasi-static tensile properties, adopted for preparing the fatigue tests; the detection of the stress–strain level for damage initiation and development and observation of the damage progression during loading;

– the fatigue life diagram and some damage metric representations; the damage observation and evolution for different stress levels;

– the residual quasi-static tensile mechanical properties to assess the effect of the imparted fatigue damage on the postfatigue performance and damage mechanisms.

The discussed investigations are based on the same experimental procedure and had several common testing devices and measurement techniques. They are first presented in the following section.

The understandings on fatigue of textile composites collected in this chapter have benefited from years of collaboration with A.E. Bogdanovich, whose depth of knowledge and vast experience of textile reinforcements for composites enriched our research [CAR 10a, CAR 13, IVA 09a, KAR 11, LOM 09].

1.2. Experimental methodologies

The results and discussion in this part of the book are focused only on quasi-static tensile and tension–tension cyclic loading, acknowledging that they only provide partial insights about the behavior of the considered materials. However, these loading conditions are considered to be of primary importance and the first step for any investigation related to the fatigue behavior of composite materials [HAR 03, ANO 15].

This section summarizes the main features of the experimental setup common to the investigations presented in the chapter, including the methodology and devices for damage detection and observation. Peculiarities are mentioned in the following sections dedicated to each of the considered composite materials.

Typically, a prismatic shape of the tensile specimen with overall length 250 mm, gauge length 150 mm and width 25 mm is used. This is prescribed by a number of international standards, for example ISO 527-4, ISO 13003, ASTM D3039, ASTM D3479, JIS K 7083. An appropriate preparation of the clamping zones of the specimen is mandatory to avoid undesirable failure mainly in fatigue loading [DEB 08, DEB 09]. Aluminum or multilayers glass reinforced plastic tabs and epoxy adhesive were used in the experimental investigations considered.

Pre- and postfatigue quasi-static tensile tests were conducted assuming a crosshead speed of 1 or 2 mm/min. Grips pressure turned out to be a critical aspect, mainly for cyclic loading. An appropriate choice avoids sliding of the specimen under increasing load, or failure inside or close to tabs induced by excessive stress concentration. As a good practice, failure of specimens at a distance of less 2 cm from tabs were discarded, both in quasi-static and fatigue tests, to exclude the influence of tabs and grip zones on the results.

The common features of the load controlled fatigue tension–tension tests were as follows: constant stress amplitude, sinusoidal wave-form of loading, ratio of the minimum to the maximum stress in the cycle $R = 0.1$, frequency in the range of 1–10 Hz depending on the material and the load level. This frequency range was motivated by the influence on the mechanical response of the loading frequency and the specimen heating. Preliminary investigations were conducted on both principal in-plane directions of the composites PWG (section 1.3) and 3DW (section 1.4), with dynamic mechanical analysis bending tests, assuming frequency in the range of 0.1–80 Hz and temperature ranging from room temperature to 80 °C. Below 40° C, the mechanical response of the materials did not have relevant variation for loading frequency between 1 and 30 Hz.

The damage development during quasi-static tension was monitored, using acoustic emission (AE) recording equipment and an optical system for full-field surface strain evaluation.

The fast release of strain energy due to an induced damage in a material can be detected as an AE. Following the procedure adopted in [LOM 08a], two AE sensors were situated at the boundaries of the gauge length, 15 mm from each tab, to record the AE up to 70–80% of the average ultimate tensile strength. This load level was set to avoid damage of the acoustic equipment at the specimen failure. The AE system (AMSY-5; Vallen Systems GmbH) was adopted for calibration of sensors, appropriate filtering and recording of several features of the AE events. AE analysis was used for identification of damage thresholds, i.e. stress (or strain) levels that manifest different stages of damage development. Typically, in textile composites, two main thresholds are recognized [KOI 09, LOM 09]: the first corresponds to the onset of the transverse cracking; the second corresponds to the onset of local delamination and the formation of extended transverse cracks. These damage thresholds, detected during quasi-static loading, have a correlation with the fatigue life of the composite, as discussed in detail in Chapter 2. The damage thresholds can be identified using the cumulative energy of the AE events, as proposed in [LOM 08a] and [TRU 05]. The procedure of the damage thresholds identification, as adopted for the materials presented in this chapter, is also detailed in [LOM 09].

Digital image correlation (DIC) is a contactless technique that offers qualitative and quantitative information on the heterogeneous deformation of an object surface. It provides a full-field displacement over a 2D or 3D surface, by comparing images before and after deformation [SUT 09]. For a better image correlation, the specimen surface is speckled with a random pattern of black paint over a substrate of white paint (or vice versa). Images, taken by a digital camera during tensile loading of composite specimen, are compared by a dedicated software (for the investigations in this chapter, VIC-2D;

Correlated Solutions Inc.) to either the initial image or the previous image determining local displacement and computing strain components. The local strain can be averaged over the area of interest to give the global strain, as obtained with an extensometer. The architecture of textile composites can generate significant variation of the local strain distribution. The modification of the full-field strain, increasing the load, allows detecting the global stress or strain levels for which local concentrations of strain identify damage onset or evolution on the observed surface [LOM 08b]. This is an important information to be linked to the thresholds obtained by the AE recording and the direct damage observations with different methodologies.

The selection of the methodology to observe damage in composite depends on the material components. Glass fiber textiles embedded in an epoxy matrix create materials transparent enough to clearly see crack development with backlit specimens using a bright transmitting light (as in [IVA 09a] and [LOM 09]). Images taken during the test allow continuous monitoring of the onset and evolution of the damage for the complete loading. The observations do not involve stopping the test at different load levels and were adopted either during quasi-static or fatigue tests. Backlit observation and DIC strain mapping were performed simultaneously on the same specimen surface to obtain a direct correlation of the strain concentrations and the damage mode.

Damage imparted in the carbon fiber reinforced composites was observed by optical microscope or X-ray micro-computed tomography (micro-CT). Particularly interesting are the observations with the micro-CT system [GAR 16, YU 15]. The connections of many X-ray images, taken from different angles, produce cross-sectional images of a portion of the scanned specimen allowing inside views without cutting. This non-destructive technique enables observation of damage with very high magnifications detecting matrix cracks within the yarns and at the fiber–matrix interface. The optical microscope and micro-CT observations were used to detect damage during quasi-static and fatigue tests at predefined load levels or number of cycles. However, those methodologies do not allow for a continuous monitoring of the damage evolution: interruption of the test is required to place the specimen, or part of, in the observation device.

1.3. Fatigue behavior and damage evolution in 2D E-glass plain weave textile-reinforced epoxy composite

The laminated composite is reinforced with four plies of plain weave E-glass textile with a total density of 3,260 g/m^2. The 2D woven textile was produced on a Dornier weaving machine. The plain weave textile consists of 2275 tex PPG Hybon 2022 E-glass roving both in warp and weft directions. The insertion densities are 1.95 ends/cm in warp and 1.60 picks/cm in weft direction (Figure 1.1). Resulting from this, the fabric is slightly unbalanced with respect to the total fiber volume content in the warp and weft directions. To have the overall fiber balance in the composite laminate, the fabric layer orientation 0/90/90/0° was assumed laying the four-ply preform. Due to the balanced warp- and weft-directional fiber content, characterization of the PWG composite was performed only in the warp direction.

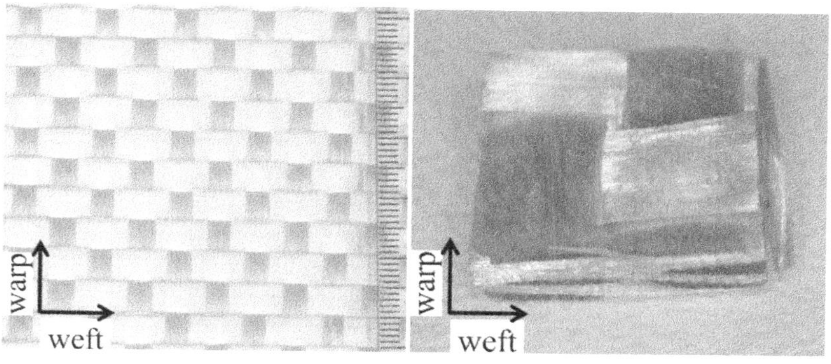

Figure 1.1. *E-glass plain weave preform*

Composite panels (PWG in the following) were fabricated using Dow Derakane 8084 Epoxy-Vinyl Ester resin (glass transition temperature 115 °C) in a vacuum bag using Vacuum Assisted Resin Transfer Molding (VARTM) method at room temperature. The composite had thickness of

2.57 ± 0.09 mm and fiber volume fraction 54.42 ± 0.65% [CAR 10a]. Similar material, possessing a slightly different fiber volume fraction, was considered for a comprehensive experimental study of the quasi-static tensile mechanical properties and progression of damage in [LOM 09] and [IVA 09a].

The experimental setup for tensile quasi-static and fatigue investigation of the PWG is as described in section 1.2, including AE and DIC. The damage onset and development observation was conducted during loading with images of backlit specimens.

1.3.1. *Quasi-static tensile behavior and damage observation*

The results of quasi-static in-plane tensile loading up to failure are summarized in Figure 1.2 with typical stress–strain curves and in Table 1.1 in terms of the Young's modulus (E), ultimate stress (σ_u) and ultimate strain (ε_u). They are in agreement with data in [LOM 09]. Figure 1.2 shows a slight nonlinearity at strain of approximately 0.3–0.5%, which highlights a possible damage generation.

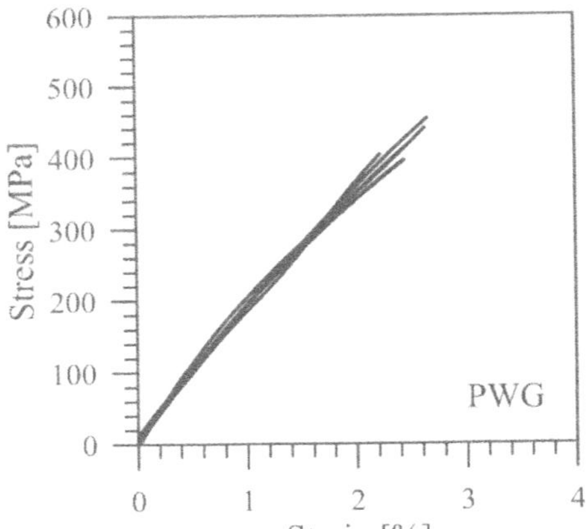

Figure 1.2. *Plain weave E-glass composite. Quasi-static tensile tests stress versus strain curves*

E (GPa)	σ_u (MPa)	ε_u (%)
24.7 ± 1.51	427 ± 23	2.45 ± 0.18

Table 1.1. *Plain weave E-glass composite. Quasi-static tensile mechanical properties. Average and standard deviation of six specimens*

AEs registration during quasi-static tensile loading gave the first understanding of the load levels for damage initiation and evolution, as mentioned in section 1.2 and explained in [LOM 08a]. The AE cumulative energy curve provides the characteristic thresholds of the damage (see Figure 1.3), which are supposed at the change in the rate of AE events accumulation. Abrupt slope variations ("knees") of the cumulative curve point out a switch to another damage mechanism. Thresholds are typically represented as strain levels. Here, stress levels are adopted to have the direct connection to the load levels of tensile cyclic tests. For the sake of completeness, both stress and strain levels are detailed in Figure 1.4.

Low energy events ($\approx 10^2$) start to occur with low frequency at a certain load threshold σ_{min} (ε_{min}). Then, AE events increase sharply in frequency and energy (10^4–10^7). The first abrupt increases in the slope of the AE cumulative energy curve corresponds to the first damage threshold level σ_1 (ε_1) (Figure 1.3). This is considered to be the damage initiation threshold. The damage affects the stiffness of the material resulting in nonlinearity of the stress–strain curve, as observed for the PWG composite in Figure 1.2. The estimated value of ε_1 is in the range mentioned above. Increasing the level of the applied load, a second "knee" on the AE cumulative energy curve appears. It corresponds to the second damage threshold σ_2 (ε_2), which indicates an evolution of the damage mode as observed with strain mapping and backlit images.

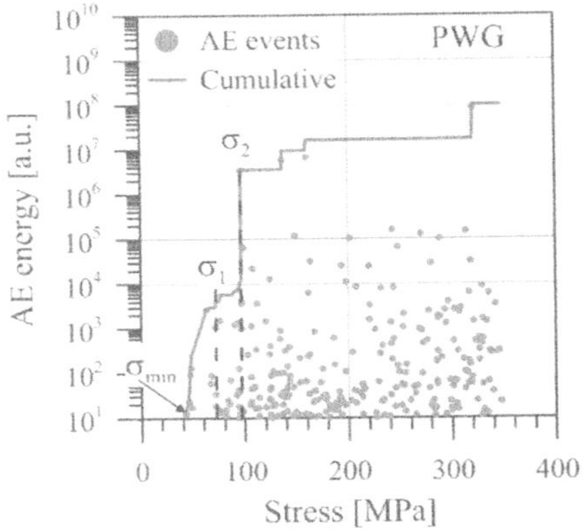

Figure 1.3. *Plain weave E-glass composite. Quasi-static tensile tests: representative AE energy versus stress diagram*

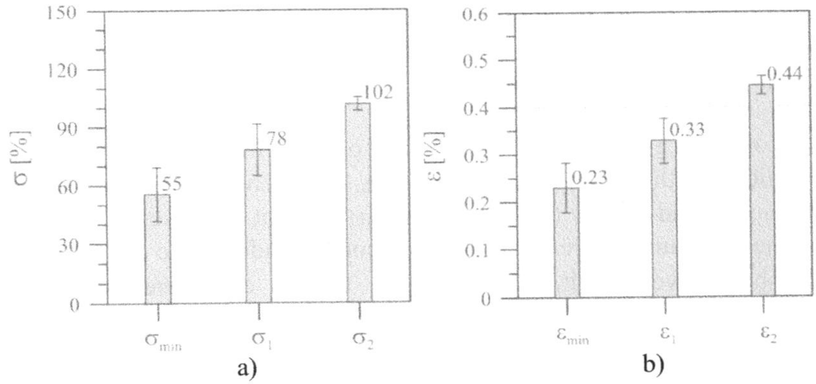

Figure 1.4. *Plain weave E-glass composite. Quasi-static tensile a) stress and b) strain thresholds. Average and standard deviation of six specimens*

At the load level corresponding to σ_{min} (ε_{min}), the PWG composite experienced onset of some matrix cracks, transverse to the load direction, (Figure 1.5), which do not influence the material stiffness. In the same zone, strain bands of higher values are visible. When the load increases up the level of the threshold σ_1 (ε_1), the preexisting cracks increase in length and new short transverse cracks appear (Figure 1.5). At the load level corresponding to σ_2 (ε_2), the same damage mode evolves with a faster growth in length and multiplication of cracks. The random distribution of cracks over the observed surface is mainly connected to the random nesting of the layers resulting in an irregular pattern of strain map (Figure 1.5). As suggested in [IVA 09a], those transverse cracks could be located within the weft yarns.

The main understandings, for this material, from the correlation of the AE recordings and the damage observation are as follows: the first AE threshold (σ_1/ε_1) is connected to the onset of new transverse cracks and the increase in length of the few previously created; the second AE threshold (σ_2/ε_2) is connected to a faster growth in length and multiplication of transverse cracks.

For the sake of completeness, after removing the AE sensors (for a stress level above σ_2), the damage evolution was observed with backlit observations.

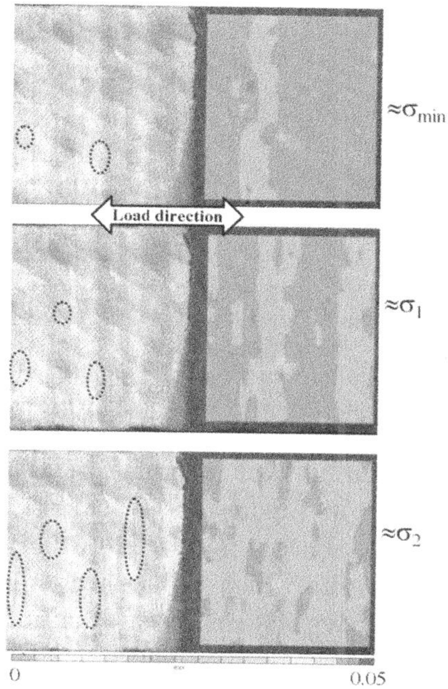

Figure 1.5. *Plain weave E-glass composite. Quasi-static tensile tests. Damage observation at the stress thresholds: backlit image (vertical side covers the entire width) and map of the strain component in the load direction*

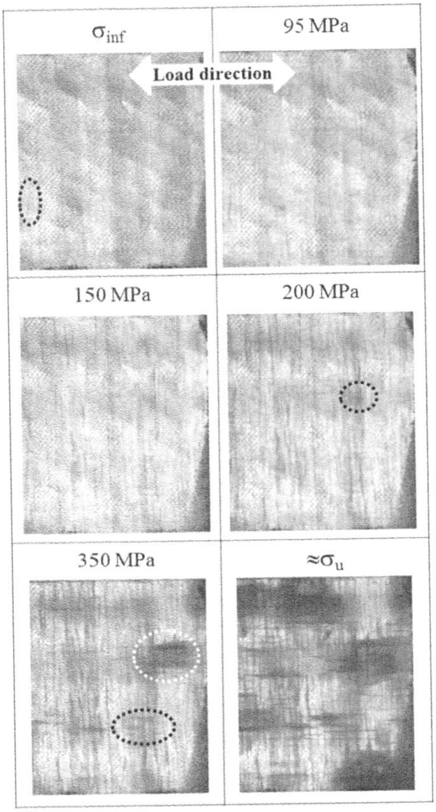

Figure 1.6. *Plain weave E-glass composite. Quasi-static tensile tests. Damage at different stress levels: backlit images (vertical side covers the entire width)*

Images collected during the complete loading of the same specimen are compared in Figure 1.6. The first load level (σ_{inf} = 60 MPa in between σ_{min} and σ_1) is related to the lower fatigue load level (see section 1.3.2) for which failure was not recorded after a predefined number of cycles. This is to have an understanding on the influence of the initial damage on the long fatigue life of the material. For this load level, onset of few transverse cracks was visible in agreement with Figure 1.5.

For a stress level of 95 MPa, the PWG has developed diffuse transverse cracks in the weft direction. This is expected since the load level is close to the threshold σ_2 (Figure 1.4). The transverse cracks multiply and become longer, with the length sometimes spanning over some unit cells (Figure 1.6). Transverse cracks have lengths of several unit cells, when loading reaches 150 MPa. The location of these cracks seems random in different plies of the PWG laminate. The load of 200 MPa generates a crack density close to saturation and onset of longitudinal cracks in warp yarns starts to be visible. On approaching the ultimate failure load, onset of delamination occurs in PWG (see load level 350 MPa, Figure 1.6). Finally, splitting of the weft yarns and generation of large delaminated regions between the layers appear when load reach the ultimate level (σ_u).

In summary, the main damage modes observed during quasi-static loading of PWG composite are as follows: transverse cracks, longitudinal crack and large regions of delamination between the layers.

1.3.2. *Fatigue life and damage metrics*

The wide range of stress levels (nine) allowed depicting a fatigue life diagram (maximum stress in the cycle σ_{max} vs. number of cycles to failure). The semilogarithmic representation, in Figure 1.7(a), has maximum stress levels normalized to the quasi-static strength for a better comparison to other materials with different values of the ultimate stress.

The first result from the diagram is the stress level for which damage does not progress significantly, and the specimen does not fail after a predefined number of cycles, here set to 5 million. This stress level, named in the following "fatigue limit" σ_{inf}, was 60 MPa for the PWG composite. The "fatigue limit" does not imply resistance of the materials for any large number of cycles, but for an accepted large number of cycles, here 5 million, the initial imparted damage (see section 1.3.1) does not have an evolution such that the stress level leads to complete failure. Most often, as discussed in Chapter 2, the conventional number of cycles for the "fatigue limit" is selected as 1 million ("operational definition").

This result gives the first connection between the AE damage thresholds, recorded during quasi-static loading, and the "fatigue limit" of the textile composite, which is discussed in detail in Chapter 2.

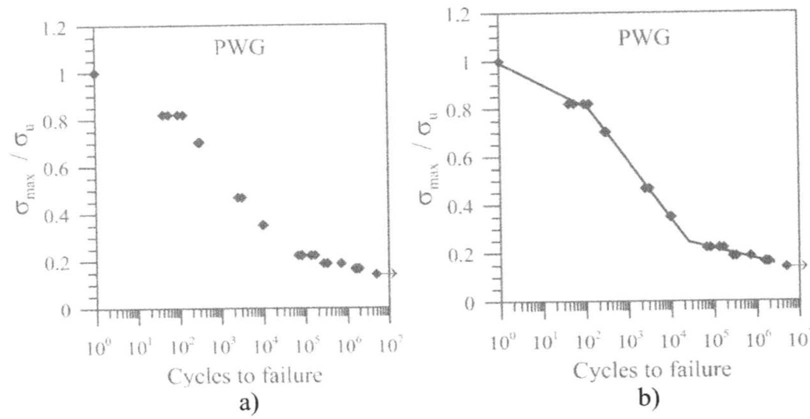

Figure 1.7. *Plain weave E-glass composite. a) Fatigue life diagram: normalized stress (maximum stress in the cycle σ_{max}/quasi-static tensile strength σ_u) versus number of cycles to failure. b) Fitting of the fatigue life diagram; "→", no failure*

The second observation from the diagram arises considering the semilogarithmic fitting of the experimental results (Figure 1.7(b)). An appropriate fitting highlights three linear segments with a correlation coefficient R^2 higher than 0.95. The three detected distinct stress ranges, with different segment slopes, are consequence of the different evolutions and modes of the fatigue damage, as detailed in section 1.3.3. The observed separation has some analogies with the three regions of the fatigue life diagram previously studied and theorized for unidirectional composites in [TAL 81, TAL 85] (see also for multidirectional laminates [TAL 08, TAL 15]). A critical discussion on this distinction is given in Chapter 2.

Some differences of the damage evolution for load levels belonging to the three stress ranges can be deduced considering some empirical metrics [GAG 06]. The main damage metric in the literature is the stiffness degradation [GUD 10]. In this chapter, two different damage metrics are used: the slope of the segment passing through the maximum and minimum points of the stress–displacement cycle curve ("cycle slope"); the energy dissipated in the cyclic loading, i.e. the area underneath the stress–displacement cycle curve ("cycle dissipation"). For clarity's sake, the cycle slope does not correspond to the stiffness of the material because it is not separated from the effects of compliance of the testing machine, but they provide the same qualitative information, as observed in [CAR 16].

The diagrams in Figure 1.8 highlight typical three stage curves [GAG 06]. The ratio to the initial value of cycle slope and dissipation is adopted to have a direct comparison with other materials. Initial stage with a rapid decrease in the cycle slope and a rapid increase in the cycle dissipation demonstrates a fast development of the damage until almost 15% of the fatigue life. The initial stage is less visible for the highest considered stress level as the damage continuously spreads over the material, which is the load level in the range closer to the quasi-static strength. In the second stage, the composite has a slowest diffusion of damage up to almost 90% of the fatigue loading. Depending on the stress level, the decrease in the cycle slope (increase in the dissipation) is proportional to the load level, meaning, as expected, slow diffusion of damage for σ_{max} = 60 MPa and fast diffusion for the highest one (σ_{max} = 300 MPa). The third and final stage (Figure 1.8) indicates a rapid decrease in the slope (increase in the dissipation) as a consequence of the rapid spread of the damage up to failure. This is not visible for σ_{max} = 60 MPa; the load level for which failure did not occur after 5 million cycles.

The three cyclic stress levels initially create three different damage patterns as observed during the quasi-static loading (section 1.3.1). For the stress level in the lowest fatigue stress range (i.e. σ_{inf}, just below the first AE threshold), the initial damage is mainly onset of few transverse cracks (Figure 1.6). The stress level (σ_{max} = 150 MPa, just above the second AE threshold), in the second fatigue stress range, creates diffuse transverse cracks in weft direction. The highest stress level (σ_{max} = 300 MPa), in the fatigue stress range closest to the quasi-static strength, is responsible for longitudinal cracks in warp yarns and onset of delamination. The evolution of these initial damage patterns during cyclic loading was observed with backlit images as detailed in the following section.

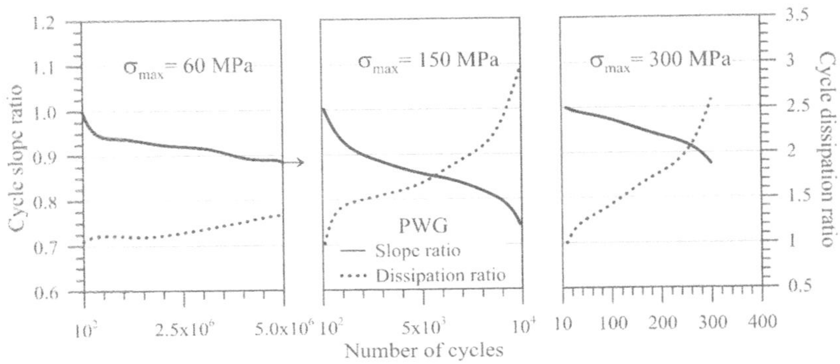

Figure 1.8. *Plain weave E-glass composite. Comparison of the cycle slope ratio and cycle dissipation ratio for some maximum stress levels. "→", no failure*

1.3.3. *Fatigue damage observation and evolution*

The evolution of the damage, observed by backlit images recorded during cyclic loading without tests intervals, was considered for two stress levels: the level for which failure did not occur after 5 million cycles (σ_{inf} = 60 MPa), belonging to the lowest fatigue stress range; the stress level σ_{max} = 200 MPa, in the second fatigue stress range, which imparts, as quasi-static load, diffuse transverse cracks in weft direction and onset of longitudinal cracks (Figure 1.6).

The few transverse cracks in the first cycle of the stress level σ_{inf}, as observed for quasi-static loading, develop rapidly in length and number in the first part of the cyclic loading (see picture for 10^4 cycles in Figure 1.9 right). Their density becomes relevant after 10^5 cycles with the very first initiation of longitudinal cracks. The damage developed in this part of the fatigue loading correspond to the initial stage of rapid decrease in the cycle slope (increase in dissipation) visible in Figure 1.8 for this stress level.

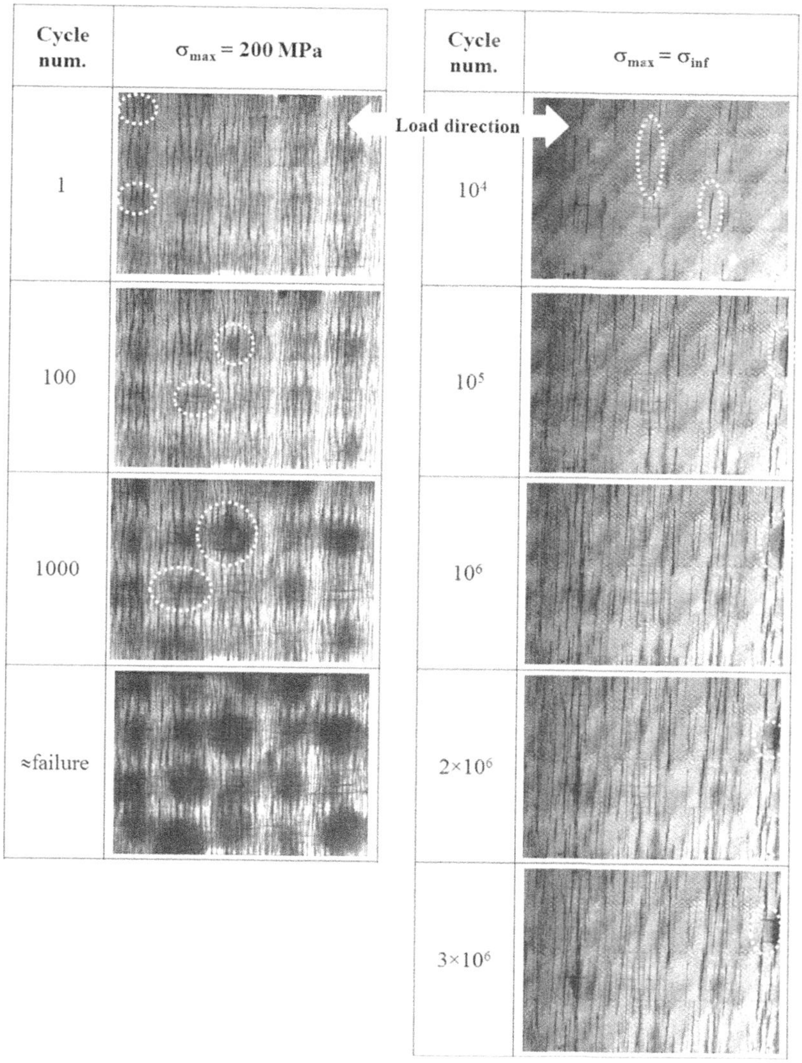

Figure 1.9. *Plain weave E-glass composite. Fatigue test maximum stress level 200 MPa (left column) and σ_{inf} (right column). Damage at different cycles: backlit images (vertical side covers the entire width)*

Protracting cycling further up to 3 million cycles, the transverse cracks have a gradual increase in length and width with almost stable pattern. Moreover, the length of the longitudinal cracks gradually increases as well, remaining localized in a zone where the early stage of delamination occurred. The consequence of this evolution of the fatigue damage is the slow decrease in the cycle slope (increase in the dissipation) observed in Figure 1.8.

The connection of the damage observations and the damage metrics suggest that a fatigue load level below the first AE damage thresholds is characterized by a damage mode involving transverse cracks increasing in length, width and density. An initiation of a second damage mode, namely longitudinal cracks, is visible only in the late stage after 1 million cycles. But, both the main and the secondary damage mode do not lead to failure of the material after 5 million cycles. This suggests that this fatigue load level can be considered as the "fatigue limit" in the extent defined and discussed in Chapter 2.

The evolution of the damage for the load level σ_{inf} has some analogies with damage observed for quasi-static tensile tests (section 1.3.1). Crack initiation and propagation for an increasing number of cycles are consistent with the damage pattern observed for increasing loads in quasi-static tension up to a level between 200 and 300 MPa (see Figure 1.6). The effect of 5 million cycles at the considered stress level is a damage pattern similar to that imparted quasi-statically, increasing the load up to about half of the material tensile strength.

The damage pattern after the first cycle of fatigue loading with maximum stress of 200 MPa (Figure 1.9, left) is similar to that seen for quasi-static tension at the same stress level: diffuse transverse cracks and onset of longitudinal cracks in warp yarns (Figure 1.6). The latter damage mode has an intensive development in the initial 300 cycles (corresponding to almost 10% of the fatigue life). Further cyclic loading creates longer longitudinal cracks and onset of the new damage mode, namely delamination (see picture for 10^3 cycles in Figure 1.9, left). The onset of delamination is located at the yarn intersection zones. The delamination evolves up to almost 90% of the fatigue life, while in the remaining loading the delaminated zones interconnect and rapidly propagate, leading to failure (Figure 1.9).

For this fatigue load level, the three stages of the cycle slope (cycle dissipation) curve are related to the damage evolution as: in the first stage, saturation of the transverse cracks and intensive developments of the longitudinal cracks create the fast reduction of the slope (increase in dissipation); in the second stage, the slower reduction of cycle slope (increase in dissipation) is mainly governed by the generation of different delamination zones; the fast reduction of the cycle slope (increase in cycle dissipation), finally, depends on the diffusion and interconnection of the delaminated areas.

The sequence of damage modes observed for this fatigue load level is analogous to the evolution of the damage modes in quasi-static loading, from the load level corresponding to the max stress in the cycle (200 MPa) to the failure (see Figure 1.6).

1.3.4. *Postfatigue mechanical properties and damage observation*

The influence of the fatigue damage on the mechanical properties of the PWG composite was investigated after different numbers of cycles (1, 3 and 5 million) with the stress level of $\sigma_{inf} = 60$ MPa. The retentions of the main tensile mechanical properties are shown in Figure 1.10. The average values are of two specimens for each number of cycles. The damage developed in the first loading stage creates a fast reduction of the cycle slope (increase in the cycle dissipation) (see Figure 1.8, $\sigma_{max} = 60$ MPa) and as a result, the largest reduction of the elastic modulus (E) and strength (σ_u) as recorded after one million cycles (Figure 1.10). The elastic modulus and strength received about 87% and 68% of the prefatigue value, respectively. In the second and third stages, the cycle slope (dissipation) curve had a slight decrease (increase) resulting in an almost stable elastic modulus and a further 17% reduction of the strength. The relevant reduction of ultimate properties (σ_u and ε_u) can be related to the residual damage after fatigue, which has drastic evolution above the load level adopted in the cyclic loading, while below remains almost unchanged with a limited effect on the elastic modulus.

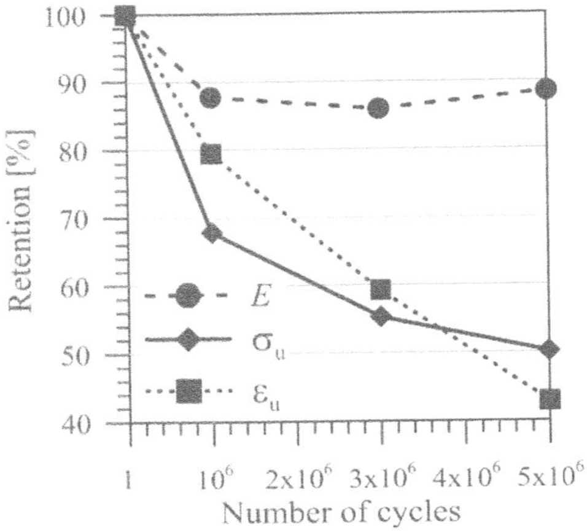

Figure 1.10. *Plain weave E-glass composite. Postfatigue quasi-static tensile tests. Average retention of the elastic modulus (E), strength (σ_U) and ultimate strain (ε_U)*

The latter comment can be directly supported by observing the AE recordings of a postfatigue tensile test after 5 million cycles in Figure 1.11(a). The number of AE events and their frequency is negligible below the fatigue load level. The number of events drastically increases once the applied load is just above the fatigue stress, namely σ_{inf}. This shows that the damage introduced during the cyclic loading remains almost unchanged as long as the static load is lower than the fatigue stress. Above this level, the damage patter rapidly evolves as AE events increase rapidly in frequency and number. The cumulative AE energy of postfatigue tensile loading is shifted to a lower stress range compared to the prefatigue counterpart (Figure 1.11(b)). This implies lower damage thresholds for the cyclically loaded PWG, meaning the imparted damage during fatigue starts develops and extends for lower load levels than for the unfatigued material.

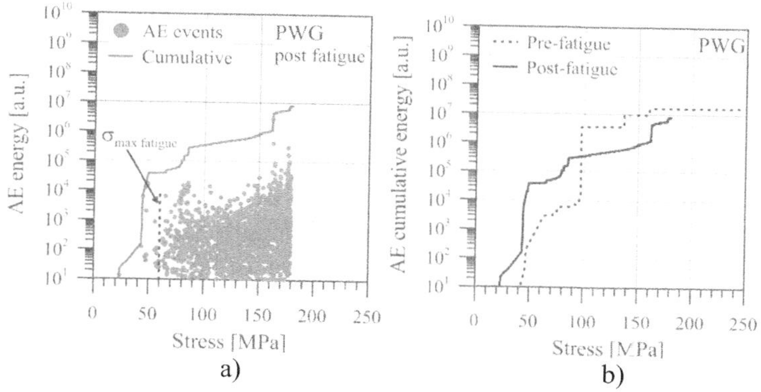

Figure 1.11. *Plain weave E-glass composite. Postfatigue quasi-static tensile test after 5 million cycles. a) Representative AE energy versus stress diagram. b) Comparison of the pre- and postfatigue cumulative energy*

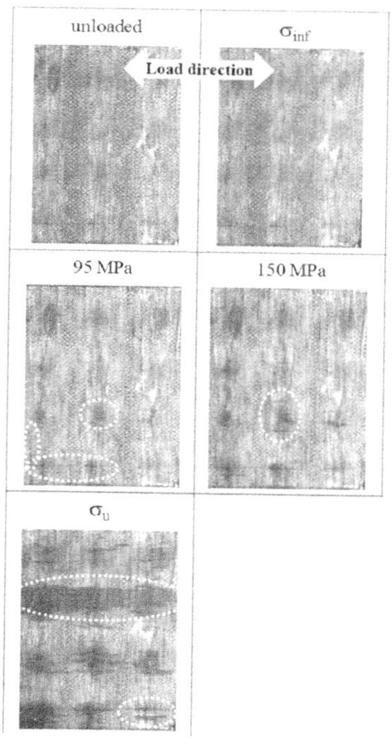

Figure 1.12. *Plain weave E-glass composite. Postfatigue quasi-static tensile test after 3 million cycles with σ_{max} = 60 MPa. Damage at different stress levels: backlit images (vertical side covers the entire width)*

The backlit images of the damage at different postfatigue load levels (Figure 1.12) confirm the AE analysis information for the specimen after 3 million cycles. The damage patterns for the unloaded state and loaded up to σ_{inf} do not show significant variations, as expected with the AE recordings. The damage pattern of the unloaded state is similar to those observed after the quasi-static load of the same level as max cyclic stress (Figure 1.6). The increase in length and multiplication of longitudinal cracks are visible for a load level of 95 MPa, while similar conditions were observed for a prefatigue quasi-static level higher than 200 MPa. Onset of delamination starts to be visible for a postfatigue load of 150 MPa and for a prefatigue stress level of about 350 MPa. In the final stage of the postfatigue test, the delamination spreads more rapidly (Figure 1.12), leading to failure with a load level σ_{u} of almost 55% of prefatigue counterpart (see Figure 1.10).

The fatigue damage accumulated has a natural evolution when quasi-statically reloaded after the imposed cyclic stress, with the same damage modes as observed in prefatigue tests.

1.4. Fatigue behavior and damage evolution in single-ply non-crimp 3D orthogonal weave E-glass reinforced epoxy composite

The reinforcement of the composite is a single-ply non-crimp 3D orthogonal weave E-glass fabric (commercialized under trademark 3WEAVE® by 3Tex Inc.). The preform has three warp and four weft layers, interlaced by through thickness (Z-directional) yarns (Figure 1.13). Table 1.2 presents some properties of the 3D preform. The fabric has a fiber amounts ratio of ~49%/~49%/~2% by volume in the warp, weft and Z fiber directions, respectively. The 3D woven textile was produced by 3TEX Inc. on a proprietary 3D weaving machine. The fiber material was PPG Hybon 2022 E-glass, as adopted for PWG.

Panels were produced using Dow Derakane 8084 epoxy-vinyl ester resin (as for PWG) and the VARTM method at room temperature. The 3D woven textile composite panels (referred to hereafter as 3DW) had thickness 2.58 ± 0.05 mm and fiber volume fraction 53.22 ± 0.6 3% [CAR 10a]. A similar 3D reinforced composite, possessing a slightly different fiber volume fraction, was experimentally investigated for quasi-static tensile in-plane stiffness and strength, and characterization of progressive damage in [LOM 09] and [IVA 09a].

In the present context, the quasi-static and fatigue tensile behavior was considered, extracting specimens with the longest side along the two in-plane principal directions of the preform, i.e. warp and weft (fill) (Figure 1.13).

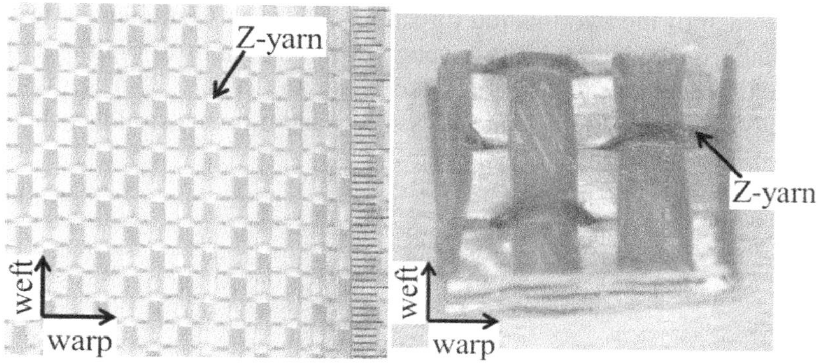

Figure 1.13. *Non-crimp 3D orthogonal weave preform*

Fabric plies		1
Density (g/m^2)		3,255
Warp layers	Insertion density (ends/cm)	2.76
	Top and bottom layer yarns (tex)	2,275
	Middle layer yarns (tex)	1,100
Weft layers	Insertion density (picks/cm)	2.64
	Yarns (tex)	1470
Z-yarns	Insertion density (ends/cm)	2.76
	Yarns (tex)	1,800

Table 1.2. *Features of the non-crimp 3D orthogonal weave E-glass perform*

The experimental investigation of the 3DW involved the same experimental setup and parameters as for PWG, described in section 1.2, including AE, DIC and damage observation during loading with images of backlit specimens.

1.4.1. *Quasi-static tensile behavior and damage observation*

For each of the two loading directions, the quasi-static tensile tests up to failure provided the stress versus strain diagrams in Figure 1.14. The main mechanical properties are listed in Table 1.3 in terms of Young's modulus (E), ultimate stress (σ_{ult}) and ultimate strain (ε_{ult}).

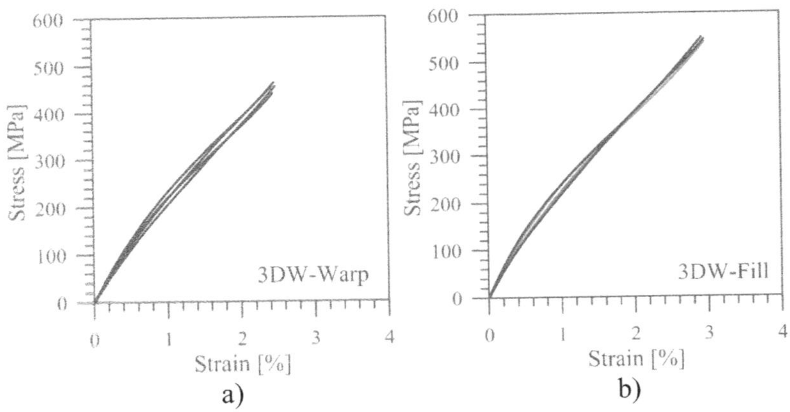

Figure 1.14. *Non-crimp 3D orthogonal weave E-glass composite. Quasi-static tensile tests stress versus strain curves: a) warp direction; b) fill direction*

	E (GPa)	σ_u (MPa)	ε_u (%)
3DW-Fill	26.3 ± 0.63	540 ± 20	2.92 ± 0.05
3DW-Warp	26.4 ± 0.76	441 ± 26	2.41 ± 0.13

Table 1.3. *Non-crimp 3D orthogonal weave E-glass composite. Quasi-static tensile mechanical properties. Average and standard deviation of six specimens*

The 3DW composite has very similar moduli in both the warp and fill directions, while a significant difference is measured for the ultimate stress and strain values (Table 1.3), as already observed in [LOM 09]. The ultimate stress of the fill-directional loaded composite is 18% higher than the warp counterpart. As suggested in [LOM 09], the difference between the ultimate stress and strain values in the warp and fill direction could be motivated with the weaving-imparted warp-directional fiber fracture, which does not affect the initial loading (similar elastic moduli), while it has a significant influence at the failure stage. The latter has some implications in the fatigue behavior as discussed in section 1.4.2.

Typical results of AE registration for both loading directions are detailed in Figure 1.15. As discussed for the PWG, the abrupt slope variations (i.e. "knees") of the cumulative curve and high rate of AE events accumulation indicate the characteristic damage thresholds [LOM 08a]. The measurements of the damage thresholds are shown in Figure 1.16 in term of stress and strain levels. The stress thresholds are adopted to connect the initial damage imparted and the fatigue behavior.

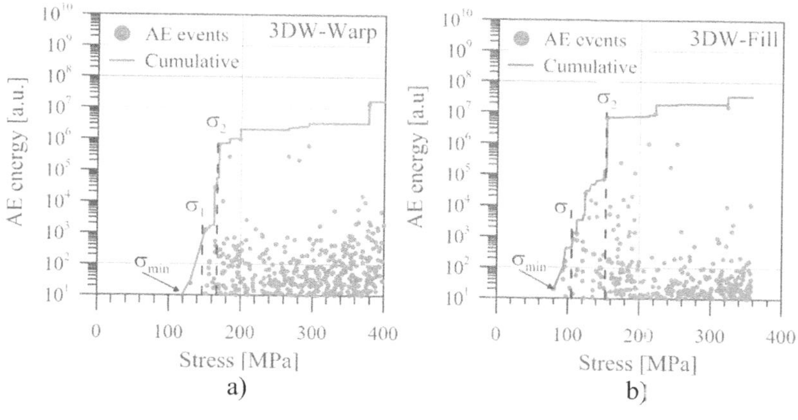

Figure 1.15. *Non-crimp 3D orthogonal weave E-glass composite. Quasi-static tensile tests: representative AE energy versus stress diagrams for a) warp direction; b) fill direction*

The low energy acoustic events start to occur infrequently at the threshold σ_{min} (ε_{min}). The first "knee" of the AE cumulative energy, corresponding to the first damage threshold level σ_1 (ε_1), is recorded in a strain range for which the damage creates nonlinearity of the stress–strain curve in both direction (Figure 1.14). The second abrupt increase in the slope of the AE cumulative energy curve, i.e. the second damage threshold σ_2 (ε_2), appears as a result of a variation of the damage mode. This was observed with strain mapping and backlit images.

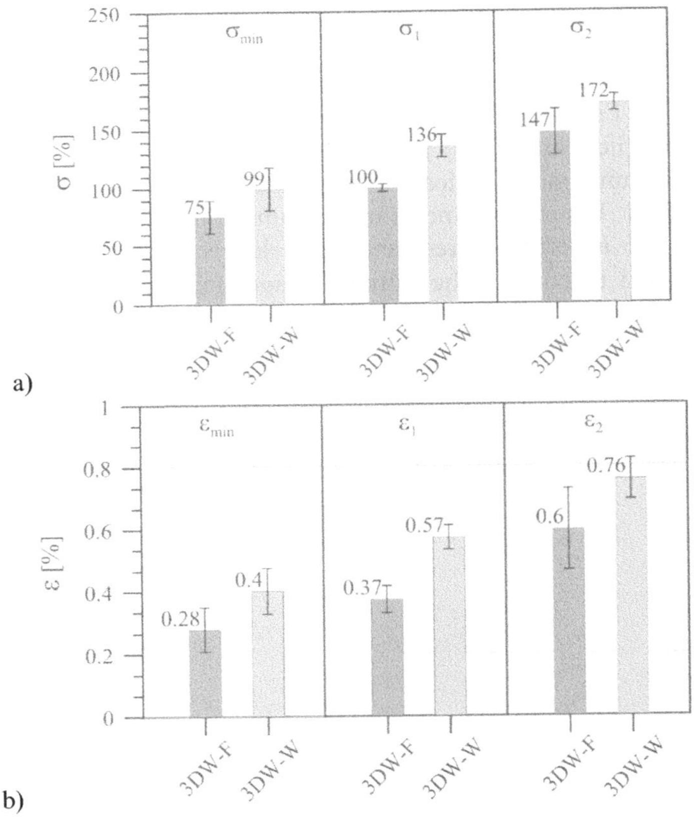

Figure 1.16. *Non-crimp 3D orthogonal weave E-glass composite. Quasi-static tensile a) stress and b) strain thresholds. Average and standard deviation of six specimens*

The comparison of the AE thresholds of the PWG (section 1.3.1) and 3DW composites (same resin and glass fiber) highlights the influence of the reinforcement architecture on damage initiation and development. The PWG has considerably lower damage thresholds. The main reasons, suggested in [LOM 09], are the absence of yarn crimp in the 3DW composite and the lowest fiber damage caused by weaving in the 3D preform [MOH 01, BOG 09].

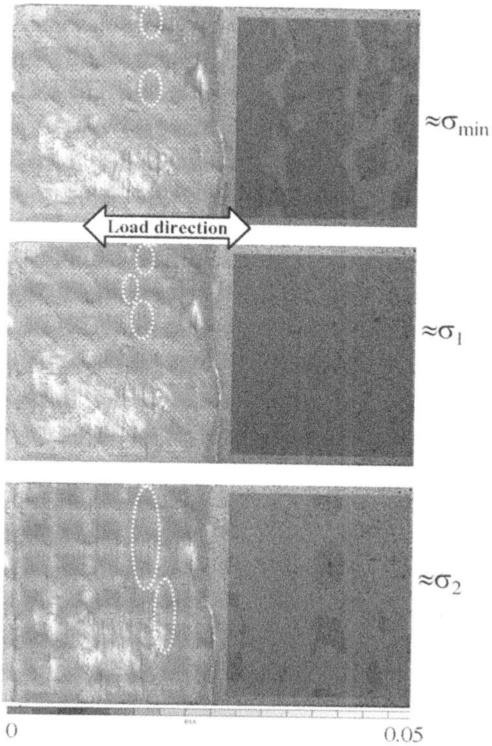

Figure 1.17. *Non-crimp 3D orthogonal weave E-glass composite. Quasi-static tensile tests fill direction. Damage observation at the stress thresholds: backlit image (vertical side covers the entire width) and map of the full-field strain component in the load direction*

In contrast to the ultimate properties, characteristic damage thresholds of the 3DW are higher for the warp direction than for the weft one (see Figure 1.16). As mentioned above and supposed in

[LOM 09], the manufacturing damage in the warp fibers [LEE 01, RUD 03] may not affect the initial loading stage, and the damage initiation thresholds could be less sensitive then the failure parameters.

To highlight the peculiarities of local damage patterns at the AE damage thresholds of the two loading directions, backlit pictures and DIC strain maps on the same specimen are considered. At threshold σ_{min}, the 3DW loaded in the fill direction had some cracks of limited size (Figure 1.17, left). These small cracks arise at the crown of Z-yarns (Z-crown), which are short surface segments of Z-yarns oriented in warp direction and laying on the fill yarns beneath them. As observed by micrographs in [IVA 09a], their exact location is at the warp yarn/matrix interface. The initial cracks are only in a few random locations that emphasizes the stochastic nature of damage initiation in the material [IVA 09a], as the correspondent strain concentration is random (Figure 1.17, right). The development of the short transverse parallel cracks previously formed has as a consequence the first damage threshold σ_1 (ε_1). Several short transverse cracks are formed, while existing ones do not grow through adjacent fill yarns.

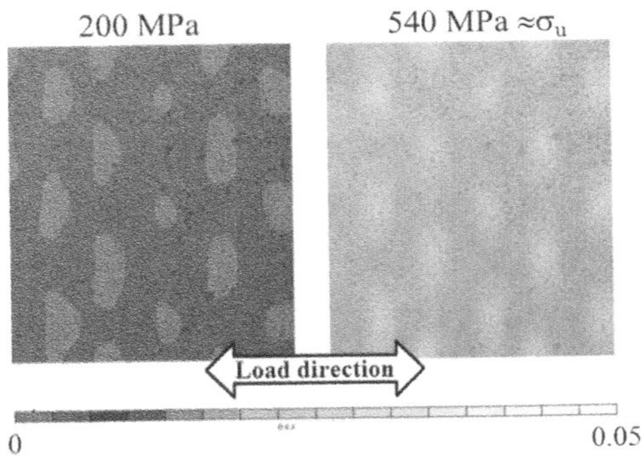

Figure 1.18. *Non-crimp 3D orthogonal weave E-glass composite. Quasi-static tensile tests fill direction. Map of the strain component in the load direction (vertical side covers the entire width)*

For the fill loading level corresponding to the second damage threshold σ_2 (ε_2), several transverse cracks close to each other are evident (Figure 1.17). The longer transverse cracks extend over the boundary of a unit cell, but their length is limited to approximately the width of two fill yarns. Once the transverse cracks at Z-crowns reach the neighboring yarns, new cracks are initiated at other Z-crowns with a fast increase in length. The tendency of multiplication rather than of propagation is visible from the analysis of strain maps for load levels higher than σ_2 (ε_2) (Figure 1.18). The fill direction loading shows a regular pattern of strain concentrations. The periodic distribution of the strain in the fill direction reflects the initiation and development of cracks at Z-crowns where higher values of surface strain are measured. These cracks develop parallel to the warp direction but their growth is hindered by fill yarns, as shown by the localized spots of strain concentration. In these spots, a pattern of close short cracks can be observed (Figure 1.17).

The 3DW loaded in the warp direction does not show visible damage, in the observed area, at the threshold σ_{min} (ε_{min}) (Figure 1.19). Damage starts to develop at stress level close to the first damage threshold σ_1 (ε_1). Diffuse short transverse cracks are initiated near the Z-crowns, as also observed in [IVA 09a], the cross-over being a natural local site of stress concentration as a consequence of the densely packed orthogonal Z and fill yarns. The transverse bands of strain concentration point out the diffuse distribution of the small cracks along the fill direction. Increasing the load up to the second damage threshold σ_2 (ε_2), those cracks spread and, after the saturation within the width of the Z-yarns, propagate and interconnect (Figure 1.19), resulting in wider transverse bands of strain concentration.

The differences of the complete damage evolution up to failure for the two loading directions, also above the second damage threshold (after removing the AE sensors), are discernible when comparing backlit images at different load levels (Figure 1.20).

The first load level σ_{inf} (60 and 55 MPa for fill and warp direction, respectively) is related to the lower fatigue load level for which failure was not recorded after a predefined number of cycles (see section 1.4.2). This is lower than the threshold σ_{min} (see Figure 1.16) and, as expected, damage is not visible for both load directions. The stress levels of 95 and 150 MPa are in the range of the first and the second damage thresholds, respectively, for weft loading, while they are in the range of σ_{inf} and σ_1 thresholds for warp direction. Therefore, the above comments on the damage patterns remain valid (see Figure 1.17 and Figure 1.19), bearing in mind that for both warp and fill loading, damage initiation is linked to the presence of Z-yarns.

Load levels higher than the second damage threshold result in transverse crack multiplication for loading in the fill direction (see stress level 200 MPa in Figure 1.20). For the same load level in the warp direction (200 MPa), transverse cracks tend to grow, interconnect and widen significantly. Additionally, cracks inside Z yarns or at the Z yarn–matrix interfaces grow with limited length.

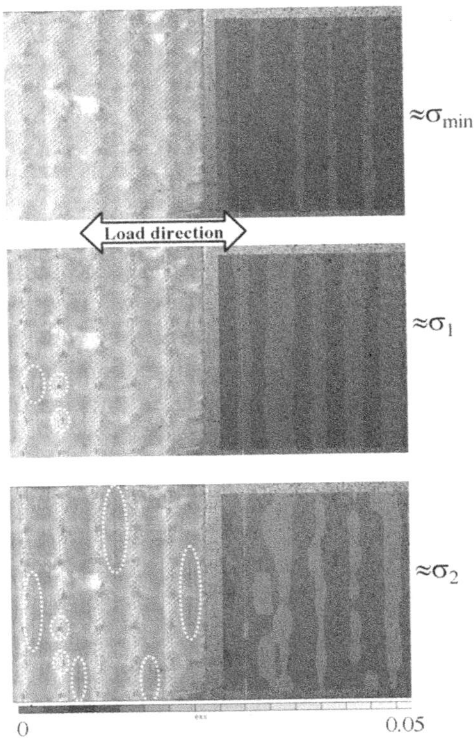

Figure 1.19. *Non-crimp 3D orthogonal weave E-glass composite. Quasi-static tensile tests warp direction. Damage observation at the stress thresholds: backlit image (vertical side covers the entire width) and map of the strain component in the load direction*

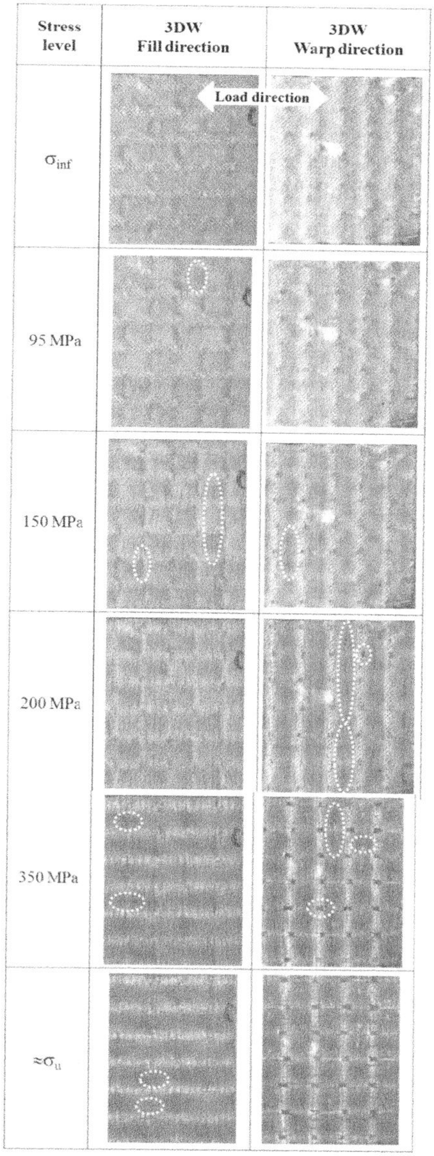

Figure 1.20. *Non-crimp 3D orthogonal weave E-glass composite. Quasi-static tensile tests. Damage at different stress levels: backlit images (vertical side covers the entire width)*

Approaching the ultimate failure load level in the warp direction, the fill yarns are almost saturated with wide transverse cracks (see stress level 350 MPa in Figure 1.20). Moreover, the cracks within Z yarns develop into local debonding at each unit cell [IVA 09a], and cracks in the fill yarns start to occur. This load (350 MPa) applied in the fill direction shows both warp and Z yarns densely populated with cracks. Cracks appear in between fill yarns in the fill direction (see Figure 1.20 for the stress level of 350 MPa), which are, according to [LOM 09] and [IVA 09a], interfacial yarn–matrix cracks. However, debonding is not observed for the fill-directional load.

Close to the failure (σ_u), the load in warp direction creates diffused local debonding around Z-yarns and rather wide longitudinal cracks. The load in the fill direction generates several cracks in the fill yarns, which are split before ultimate failure.

In summary, three main types of damage modes can be identified during quasi-static loading of 3DW composite: transverse cracks, local debonding (only for warp-directional loading) and longitudinal cracks.

1.4.2. *Fatigue life and damage metrics*

Nine load levels were considered to build the fatigue life diagram of the 3WD for both warp and fill load directions [CAR 10a]. The two semilogarithmic representations of maximum stress in the cycle σ_{max} versus number of cycles to failure are compared in Figure 1.21. As for the PWG, the normalized maximum stress to the quasi-static strength is considered for a better comparison with other materials with different ultimate stresses.

The first information from the diagram is the stress level for which the imparted damage during cyclic loading does not lead to failure of the specimen after a predefined number of cycles, here set to 5 million (as for PWG). This stress level σ_{inf} (named "fatigue limit") was 60 and

55 MPa for the 3DW, in the fill and warp direction respectively. Note again that the “fatigue limit” does not imply infinite fatigue life of the material, but it represents a load value for which failure does not happen after a conventional large enough number of cycles (see Chapter 2).

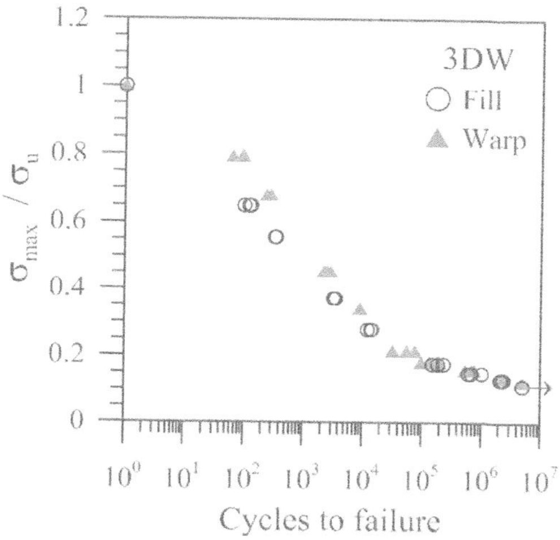

Figure 1.21. *Non-crimp 3D orthogonal weave E-glass composite. Fatigue life diagrams: normalized stress (maximum stress in the cycle σ_{max}/quasi-static tensile strength σ_u) versus number of cycles to failure. “→”, no failure*

The “fatigue limit” for both directions is lower than the AE threshold σ_{min} for quasi-static loading (see Figure 1.16). This allows correlating the “fatigue limit” to the quasi-static damage thresholds, as discussed in Chapter 2.

Comparison of the fatigue life diagrams in Figure 1.21 does not completely highlight the different fatigue responses of the 3DW composite in the principal directions. A clear understanding is observing the average fatigue life (e.g. number of cycles to failure) for different stress levels in Figure 1.22. A comprehensive statistical analysis for the cycles to failure of each stress level is detailed in [CAR 10a].

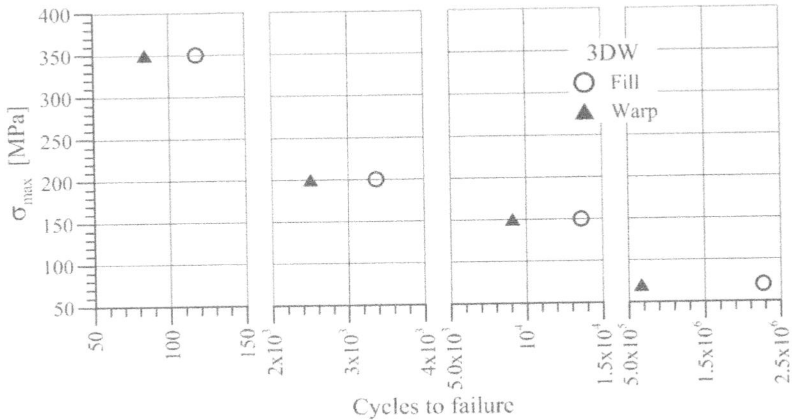

Figure 1.22. *Non-crimp 3D orthogonal weave E-glass composite. Comparison of the average fatigue life for some maximum stress levels*

The fatigue performance of the 3DW in warp and fill directions is found to be in direct correspondence with the trends observed in the static tests (section 1.4.1). The fatigue life of the 3D composite is much longer in fill than in warp direction and the difference increases on decreasing the stress level. The different fatigue performance in the warp and fill directions of the 3DW composite can be attributed to several reasons. More damage is imparted to the warp yarns than to the fill during the weaving process, as discussed for the quasi-static behavior. The local "pockets" of pure matrix are created by the Z yarns, limited to the thickness of the fill yarn layer. Despite being of small dimensions, matrix pockets play a significant role in the crack initiation of 3DW composites. Finally, as hypothesized in [CAR 10a], in warp-directional loading Z yarns are directly loaded (warp and Z yarns are parallel), while in fill loading they are not loaded directly. The cyclic frictional contact between warp and Z yarns differs between the two loading cases. The tensile loading in warp direction is applied both to warp and Z yarns, while the reduction of the specimen width due to Poisson's effect decreases the distance between adjacent Z and warp yarns, which could come into contact. Their mutual friction effect may become sufficient to significantly reduce the overall fatigue life of the material while both yarns are exposed to cyclic tensile stress.

A complete fatigue life diagram considering the semi-logarithmic fitting of the experimental data is shown in Figure 1.21. The fitting is here detailed for the warp loading direction (see the description for both principal direction in [CAR 10a]). Figure 1.23 shows the fitting of experimental results by means of three linear segments with correlation coefficient R^2 higher than 0.96. The segments subdivide the diagram in three stress ranges, which can be correlated to the different dominating damage modes imparted during the cyclic loading, as observed with backlit images in section 1.4.3. As for PWG composite, the three distinct regions of the fatigue life diagram have some analogies with the three regions related to the evolution of the fatigue damage for unidirectional and multidirectional laminates described in several works by Talreja (see, e.g., [TAL 81, TAL 08, TAL 15]). The three characteristic regions separation is discussed and compared for several textile composites in Chapter 2.

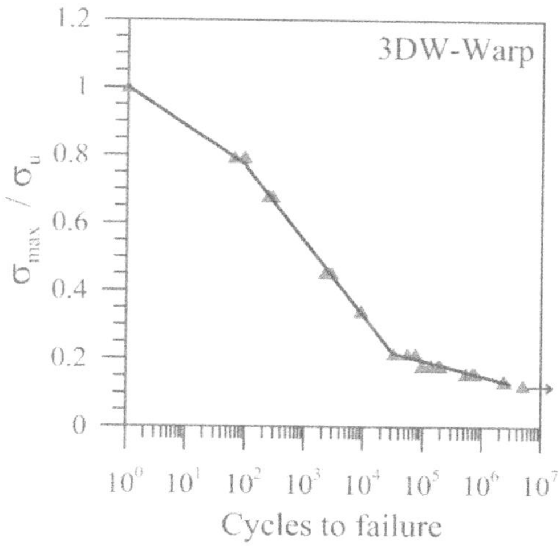

Figure 1.23. *Non-crimp 3D orthogonal weave E-glass composite. Fitting of the fatigue life diagram for warp direction loading. "→", no failure*

Some consequences of the evolution of the damage modes, distinguishing the three stress ranges of the fatigue life diagram, are visible considering the empirical metrics adopted for the PWG composite (section 1.3.2). The two metrics are: the slope of the segment passing through the maximum and minimum points of the stress–displacement cycle curve ("cycle slope"); the energy dissipated in the cyclic loading, i.e. the area underneath the stress–displacement cycle curve ("cycle dissipation"). They are represented as a function of the number of cycles, normalized to their initial values for the direct comparison of the 3DW two loading directions. The two metrics are represented, in Figure 1.24 and Figure 1.25, for three fatigue load levels, one for each stress range of the fatigue life diagram. It must be remembered that the fill load direction results in a longer fatigue life for all load levels. Moreover, the comparison for the lowest stress range of the fatigue life curve is supposed with the load level leading to the same number of cycles without failure, i.e. σ_{inf}, which is different for the two principal directions of the 3DW. As for the PWG, the diagrams show typical three stage curves [GAG 06]. This is not visible for the highest stress level for which the damage accumulates continuously, resulting in constant decreasing (or increasing) curves, similar for the two directions (diagrams for 300 MPa in Figure 1.24 and Figure 1.25).

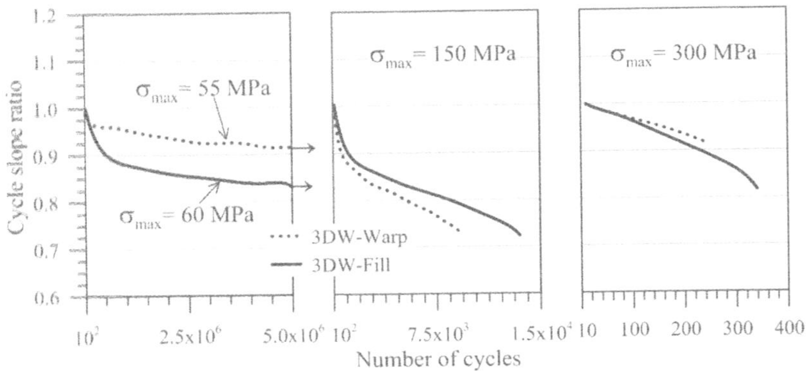

Figure 1.24. *Non-crimp 3D orthogonal weave E-glass composite. Comparison of the cycle slope ratio for some maximum stress levels. "→", no failure*

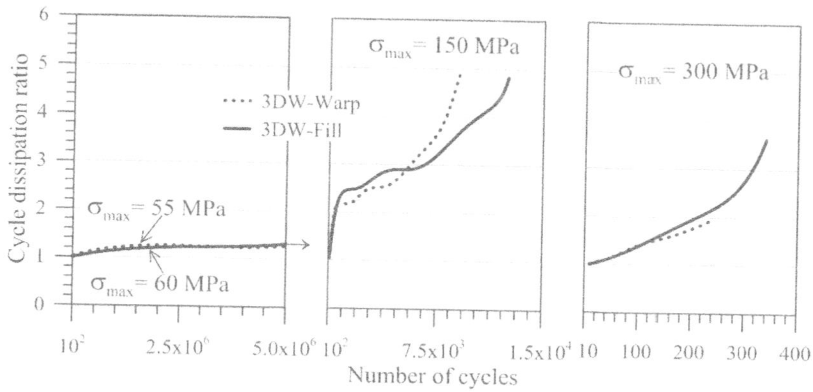

Figure 1.25. *Non-crimp 3D orthogonal weave E-glass composite. Comparison of the cycle dissipation ratio for some maximum stress levels. "→", no failure*

The first stage of the load level σ_{inf}, for a low number of cycles, has a characteristic rapid decrease in the cycle slope and a rapid increase in the cycle dissipation as a result of a fast development of the damage. This covers almost 15% and 5% of the fatigue life for fill and warp loading, respectively. That difference is an indication of the faster evolution of the damage for the fill load direction due to the higher load level than the warp direction. The second stage of the cycle slope and dissipation curves for σ_{inf} shows the slowest diffusion of damage similar for both load directions. For this load level, the third stage is barely visible because failure did not occur after 5 million cycles. The load level in the second stress range of the fatigue life diagram (see σ_{max} = 150 MPa, in Figure 1.24 and Figure 1.25) creates similar three stage curves for both principal directions. The first stage covers almost 15% of the fatigue life with a fast decrease in the cycle slope (increase in the cycle dissipation). This is a consequence of the initial quick development of the damage, which propagates more slowly in the second stage up to approximately 50% of the fatigue life. In the reaming part of the fatigue life, the rapid decrease in the slope (increase in the dissipation) indicates the rapid spread of the damage up to failure.

The three stress levels, considered for the metric of the damage evolution during cyclic loadings, impart three different damage patterns if applied quasi-statically (section 1.4.1). For the stress level in the lowest fatigue stress range (i.e. σ_{inf}, below the first AE threshold), the initial damage is not visible for both load directions (Figure 1.20). The stress level (σ_{max} = 150 MPa, in the second fatigue stress range) is close to the second and the first damage AE threshold for fill and warp loading, respectively. This load in the fill direction extends the number of transverse cracks at the Z-crowns; while, when the load is in the warp direction, several short transverse cracks initiate at the crown of Z-yarns. The quasi-static application of the stress level (σ_{max} = 300 MPa), in the highest fatigue stress range, is responsible, in the fill direction, for dense cracks distribution at the Z yarns and onset of cracks in between fill yarns; while in the warp direction this load creates wider transverse cracks and the onset of local debonding around Z-yarns. The evolution during cyclic loading of these initial damage patterns, responsible for the cycle slope and dissipation curves, is described with backlit images in the following section.

1.4.3. *Fatigue damage observation and evolution*

The direct observation of the damage development during tension–tension cycling tests is detailed for two stress levels (the same as for PWG). The first is the level for which failure did not occur after 5 million cycles, i.e. "fatigue limit" 60 MPa and 55 MPa for the 3DW fill and warp direction, respectively. They belongs to the lowest stress range of the fatigue diagrams (Figure 1.23) and are below the quasi-static AE threshold σ_{min} (ε_{min}) for which the acoustic events start to occur (Figure 1.16). The second level is 200 MPa, in the second fatigue stress range of the fatigue life diagram, above the second static damage threshold σ_2 (ε_2).

The images at the different number of cycles in Figure 1.26 and Figure 1.27 are of the same specimen for each load direction and cover its entire width.

The quasi-static loading at the σ_{inf} level did not show visible damage in the observed zone (see Figure 1.20), while after 10,000 cycles of fatigue testing certain damage was imparted (Figure 1.26).

This number of cycles generates a damage pattern similar to that observed at the first damage threshold loading in the warp direction, namely that diffuse short transverse cracks are initiated near the Z-crowns (Figure 1.19), while fill loading after the same fatigue period shows a damage distribution similar to that in between the first and the second static AE damage thresholds (Figure 1.17), that is, increase in length of cracks at Z-crowns and their multiplication. Consistent growth, interconnection and multiplication of these transverse cracks takes place up to $\approx 10^6$ cycles, with some cracks' length spanning almost the complete width (Figure 1.26). The observed diffusion of the damage in the first part of the fatigue life is responsible for the initial reduction of the cycle slope (increase in the cycle dissipation) visible in Figure 1.24 and Figure 1.25. After 1 million cycles, the damage patterns do not change significantly (only a slight increase in the transverse cracks width (Figure 1.26)) and consequently the damage metrics have slight variations. The main differences for the two loading cases under maximum stress σ_{inf} are as follows [CAR 10a]: for warp loading the existing transverse cracks increase in their length and more significantly in their width; for fill loading the transverse cracks increase in number and mutually interconnect. The growth of the length of the transverse cracks seems faster for warp than for fill loading.

The damage imparted with a fatigue max stress of 200 MPa is summarized as shown in Figure 1.27. The pattern observed after the first cycle is consistent with that seen at the same quasi-static tensile stress level (see Figure 1.20). This has a prevalent mechanism of diffuse short transverse cracks for loading in the fill direction and for the warp load direction, interconnect transverse cracks, increasing in length and width.

After approximately 100 cycles, the fatigue in the fill direction saturates the transverse cracks and initiates longitudinal cracks on Z yarn surfaces (Figure 1.27). Further cycles induce a gradual growth and multiplication of the longitudinal cracks. Approaching failure, a fast formation of transverse and longitudinal cracks inside the yarns leads to a macrosplitting and finally fiber breakage in the loading direction. This continuous evolution of the damage pattern result in the continuous reduction of the cycle slope (increase in the cycle dissipation), as for the maximum stress level in Figure 1.24 and Figure 1.25.

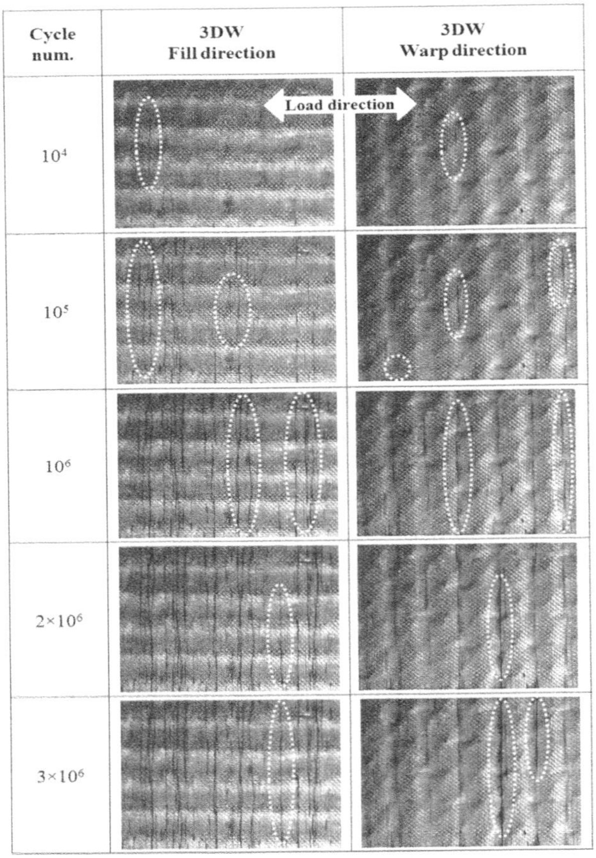

Figure 1.26. *Non-crimp 3D orthogonal weave E-glass composite. Fatigue test maximum stress level σ_{inf}. Damage at different cycles: backlit images (vertical side covers the entire width)*

The fatigue in warp direction, for the same load level, shows transversal cracks increase in length and width faster than the other direction (see 100 cycles in Figure 1.27). Onset of longitudinal cracks and local debond around Z yarns are the consequent damage modes (see 1,000 cycles). The predecessor of ultimate failure is the rapid development of these longitudinal cracks and local debonds. As for PWG, the evolution of the damage modes for this fatigue load level is consistent with that of the quasi-static loading, starting from the same load level (200 MPa) to the failure (see Figure 1.20).

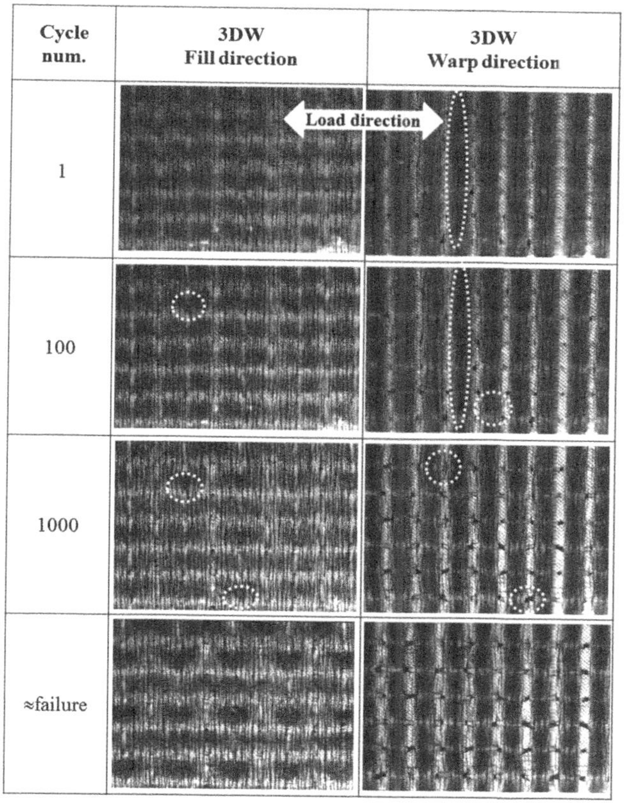

Figure 1.27. *Non-crimp 3D orthogonal weave E-glass composite. Fatigue test maximum stress level 200 MPa. Damage at different cycles: backlit images (vertical side covers the entire width)*

In summary of the observed damage evolutions, for the two fatigue load levels, the peculiar damage modes of the two lowest stress ranges of the fatigue life as shown in Figure 1.23 are: transverse cracks predominant in the stress range approaching σ_{inf} and longitudinal cracks prevalent in the second range including 200 MPa.

1.4.4. *Postfatigue mechanical properties and damage observation*

The fatigue damage causes the reduction of the quasi-static mechanical properties with respect to the unfatigued material. The mechanical properties of the 3DW in both principal directions were measured by a postfatigue tensile test of specimens, which underwent predefined number of cycles (1, 3 and 5 million cycles) with a maximum stress of σ_{inf} (60 and 55 MPa for the 3DW fill and warp direction). They included continuous recording of AE events and pictures for DIC strain calculation, and damage observation with backlit images.

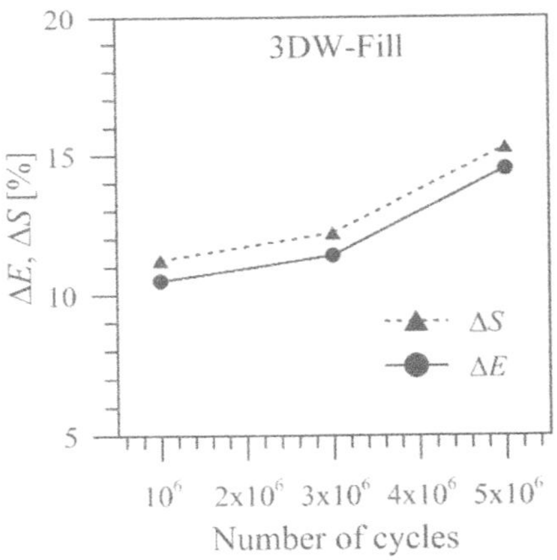

Figure 1.28. *Non-crimp 3D orthogonal weave E-glass composite. Quasi-static and fatigue tests in fill direction. Reduction of elastic modulus from pre- and postfatigue quasi-static tests (ΔE); reduction of the "cycle slope" (ΔS) after 10^6, 3×10^6 and 5×10^6 cycles, with maximum stress σ_{inf} =60 MPa*

The postfatigue elastic modulus of the material, after different numbers of cycles, allows assessing the opportunity to adopt the cycle slope as a metric of the damage imparted in the cyclic loading. The comparison, in Figure 1.28, shows the cycle slope reduction (ΔS) to the first cycle and the postfatigue static elastic modulus reduction (ΔE) to the prefatigue value of the fill loaded material. The two quantities have the same increasing trend with the number of cycles and have almost the same difference for all the number of cycles considered. The latter can be related to the unconsidered compliance of the testing machine, as mentioned in section 1.3.2. The cyclic slope, therefore, provides qualitative information (and "quantitative" variation), as for the elastic modulus, on the effect of fatigue loading on the stiffness of the materials. This confirms the conclusion in [CAR 16] comparing the cycle slope and the stiffness for each fatigue cycle.

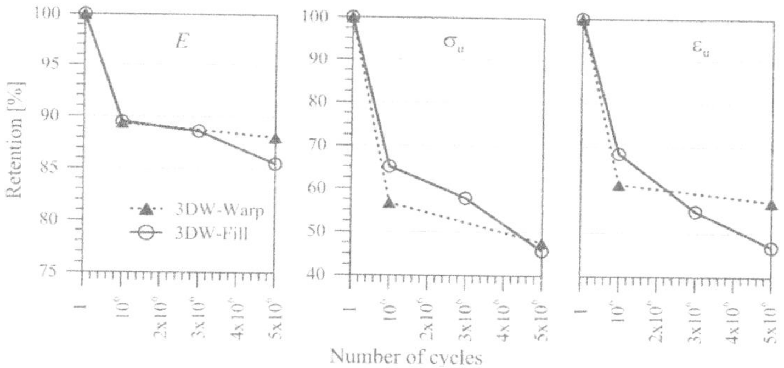

Figure 1.29. *Non-crimp 3D orthogonal weave E-glass composite. Postfatigue quasi-static tensile tests. Average retention of the elastic modulus (E), strength (σ_u) and ultimate strain (ε_u)*

An overview of the postfatigue main tensile properties, in both principal directions, in terms of retention of elastic modulus (E), ultimate stress (σ_u) and strain (ε_u) is shown in Figure 1.29. The indications of the cycle slope evolution (Figure 1.24) are now underlined with the reduction of the elastic modulus. A decrease of more than 10% of the modulus is the result of the damage imparted in the first part of the fatigue loading. In the second part, the elastic modulus has a slight reduction after 3 million cycles and remains quasi-stable (88% of the prefatigue) until the end for loading in the warp direction, while it has a further reduction of up to 86% of the unfatigued value for the fill direction. The latter is the result of the different fatigue loading level σ_{inf}, as mentioned previously. Similar information is obtained from the ultimate properties. The stress and strain at failure have the largest loss in the first part of the fatigue loading with the smaller retention for the warp direction. This could be a result of the different damage pattern (see section 1.4.3), which, despite the lower load level, has a more relevant influence on the failure properties of the warp direction. However, in the last part of the fatigue loading, the evolution of the damage due to the highest load level in the fill direction induces a slight retention of the ultimate properties.

The relative limited reduction of the elastic modulus, compared to the ultimate properties, is the result of the damage evolution during the postfatigue tensile loading. The AE events recording can justify the difference considering a postfatigue tensile test for both principal directions after 5 million cycles (Figure 1.30(a) and Figure 1.31(a)). The AE events appear consistently after the stress reaches the fatigue level σ_{inf}. This shows that the reduction of the postfatigue elastic modulus is mainly due to the fatigue damage. Further damage is not generated in the initial stage of the quasi-static loading, as expected from the comparison of the cycle slope and the elastic modulus in Figure 1.28.

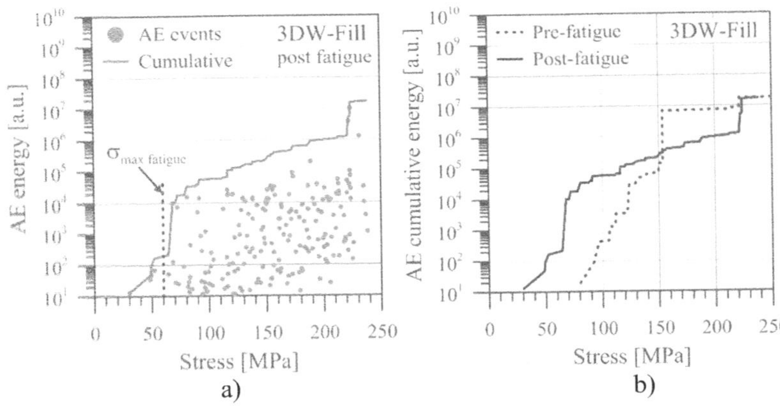

Figure 1.30. *Non-crimp 3D orthogonal weave E-glass composite. Postfatigue quasi-static tensile test fill direction after 5 million cycles. a) Representative AE energy versus stress diagram. b) Comparison of the pre- and postfatigue cumulative energy*

Above the fatigue load level, the damage starts to develop and AE events increase rapidly in number and frequency. The cumulative AE energy curve, linked to the initiation and development of the postfatigue quasi-static damage, shows an early initiation of the damage compared to the prefatigue counterpart for both loading directions (Figure 1.30(b) and Figure 1.31(b)). Each damage threshold from the postfatigue cumulative energy curve is shifted to a lower stress value, and, as a result, the damage modes are initiated and developed with smaller stress levels leading to a reduction of the failure load. The postfatigue quasi-static damage evolution at peculiar stress levels was clearly observed with backlit images.

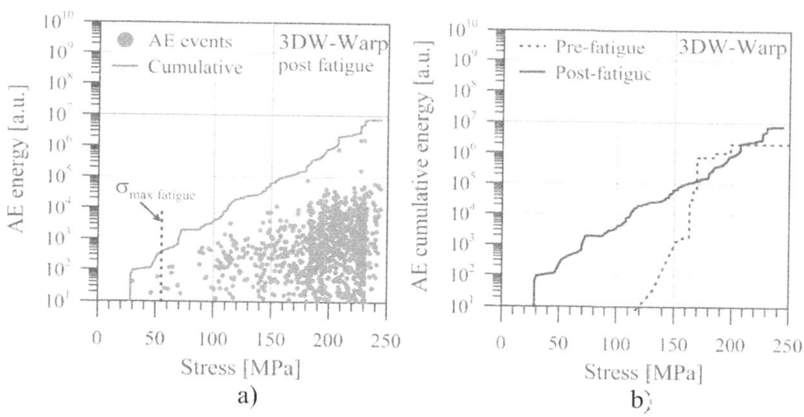

Figure 1.31. *Non-crimp 3D orthogonal weave E-glass composite. Postfatigue quasi-static tensile test warp direction after 5 million cycles. a) Representative AE energy versus stress diagram. b) Comparison of the pre- and postfatigue cumulative energy*

The backlit images in Figure 1.32 show the damage pattern after 1 million cycles (unloaded) for the two fatigued load directions. Long and wide cracks are visible for warp loading, while diffuse interconnected transverse cracks are imparted during fill fatigue loading. These are similar to the patterns presented in Figure 1.26. For the load level corresponding to σ_{inf}, the damage does not have a considerable variation, as expected for the AE events diagrams (Figure 1.30 and Figure 1.31). Above this load level, the damage evolves according to the sequence observed in the prefatigue quasi-static loading (section 1.4.1), but for reduced stress levels. Growth and further interconnection of transverse cracks and cracks in Z-yarns initiated by fatigue cycling can already be observed for an applied stress of 95 MPa (Figure 1.32); while at the analogous prefatigue stress level damage only had onset of transverse cracks of limited size (Figure 1.17 and Figure 1.19). Saturation of the transverse cracks appears for fill loading at 150 MPa, while for warp direction onset of longitudinal cracks from the Z-crowns are visible. The same fill load level of the prefatigue test creates only several transverse cracks spanning a few unit cells, while in warp loading diffuse short transverse cracks are initiated near the Z-crowns. Finally, the early appearance of longitudinal cracks and local debonding around Z-yarns (Figure 1.32) leads to faster failure for almost 65% in fill and 57% in warp direction of the prefatigue quasi-static ultimate stress (Figure 1.29).

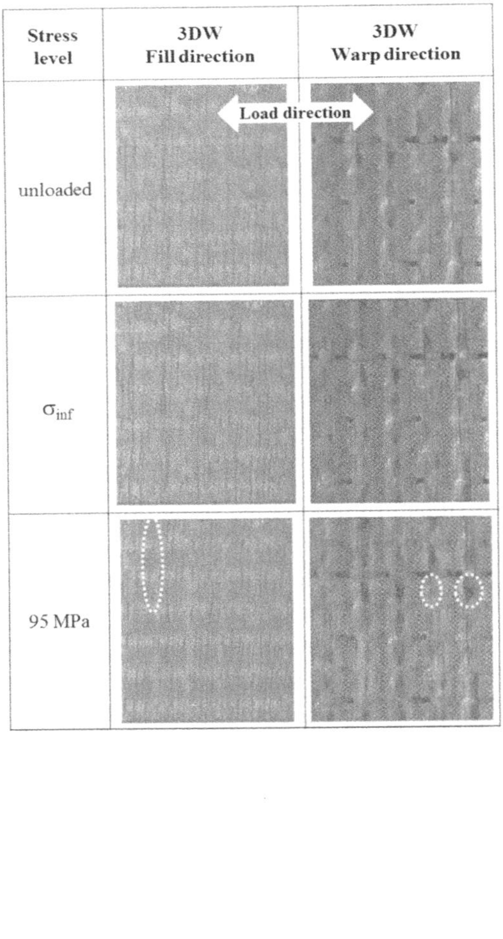

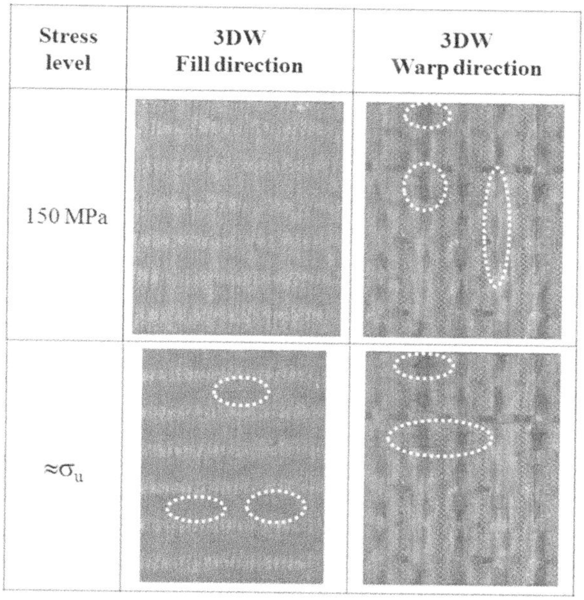

Figure 1.32. *Non-crimp 3D orthogonal weave E-glass composite. Postfatigue quasi-static tensile test after 1 million cycles with σ_{max} = 60 MPa. Damage at different stress levels: backlit images (vertical side covers the entire width)*

1.5. Fatigue behavior and damage evolution in 3D rotary braided carbon reinforced epoxy composite

The composite is based on 3D braided carbon fiber preforms (commercialized under trademark 3BRAID® by 3Tex Inc.). It is produced on the 144-horngear and 576-carrier 3D rotary braiding machine described in [MUN 02] and [MUN 07]. The braided yarns were Toho Tenax HTS 40 F13 12K carbon. The machine was used in a set-up to have the desired width of the test specimen ≈25.4 mm, and thus avoid final machining of the composite. The manufacturing of 3D braided preform with rectangular cross-section involved the total of 24 horngears (two rows of 12 horngears in each) and 96 fiber carriers. Axial yarns were not used. The adopted 3D braided preform had the architecture schematically shown in Figure 1.33. The linear weight of the preform was 78.8 ± 0.5 g/m.

The composite (referred to hereafter as 3DB) in the form of rectangular cross-section bars was fabricated at 3TEX in a pressure mold with open ends, also adopted for several other cross-sectional shape 3D braided composite structural elements [MUN 04]. The composite was produced with a room temperature cure epoxy resin West System 105 and 209 hardener. The preform was first soaked in a resin bath and then placed in the pressure mold. The resin impregnated preform was kept in the mold for at least 12 h to cure under constant pressure. Then, the fully cured composite in the form of rectangular bars of length 50 cm and width 25.4 mm was released from the mold. The experimentally determined fiber volume fraction was 55.6 ± 0.4% and the average thickness was 3.21 mm. The void content of 1.5% was estimated by considering the dry fabric and the finished composite linear weights, and knowing the fiber and resin densities. The angle between the braiding yarns and the longitudinal braiding direction was measured on the surfaces of the composite samples as 10 ± 0.5°. From geometric considerations, the "inner braid angle" was determined as ≈14° [CAR 13].

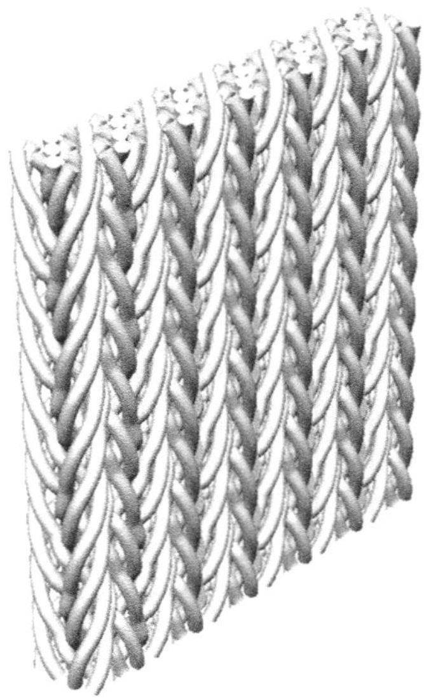

Figure 1.33. *Scheme of the interlacement in the 3D braided preform*

The tensile quasi-static and fatigue investigations, in the braiding direction, were performed with the experimental setup described in section 1.2, including AE and DIC. The damage was examined on specimens quasi-statically loaded up to failure or up to a predefined strain level with an optical microscope (Leica DM ILM inverted microscope). For fatigued specimens after a predefined number of cycles, the damage was detected by micro-CT (Philips HOMX 161 X-ray system with the AEA Tomohawk upgrade). The latter had a tube voltage of 65 kV, a current of 0.54 mA and an angular increment of 0.3 degree. The NRecon software of SkyScan NV was used for the reconstruction of the micro-CT images.

1.5.1. *Quasi-static tensile behavior and damage observation*

The quasi-static tensile tests in the braiding direction up to failure provide the complete stress versus strain curves from which the tensile static strength is extracted for setting the fatigue stress levels. The recorded stress versus strain curves show small scatter and good reproducibility of the results (see Figure 1.34) from three tests up to failure. The main measured mechanical properties of the 3DB are presented in Table 1.4.

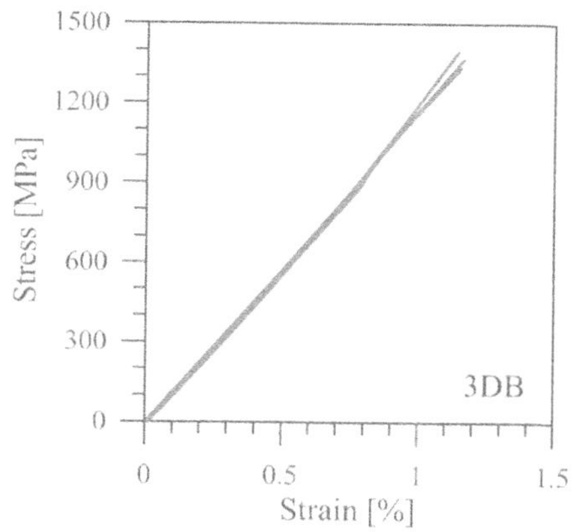

Figure 1.34. *3D braided carbon composite. Quasi-static tensile tests stress versus strain curves*

E (GPa)	σ_u (MPa)	ε_u (%)
120 ± 5	1,351 ± 17	1.15 ± 0.01

Table 1.4. *3D braided carbon composite. Quasi-static tensile mechanical properties. Average and standard deviation of five specimens*

The curves show a nonlinear behavior, evident when the tangent modulus is extracted (Figure 1.35). The stiffness increases by almost 10% with the increase in the applied strain from 0.1 to 0.8%. The stiffening is probably due to, as discussed in [CAR 13], the combination of two factors: the inherent stiffening of carbon fibers under tension, and the change in the local fiber orientation. The first is explained by a change in crystalline orientation [CUR 68, SHI 96]. The second is related to the straightening and alignment of fibers under increasing tensile load. In the 3DB composite, the braided fibers are inherently misaligned and have local crimp. Under increasing tensile load, the braided fibers tend to become aligned and straightened in the direction of loading. The damage accumulated in the 3DB composite starts to be visible from the stress–strain curves at the strain level of 0.6–0.7%; the tangent modulus reaches its maximum and then starts slowly decaying with the strain above 0.8% (Figure 1.35). In the strain interval 0.6–0.8%, the effects of fiber stiffening and developing matrix damage counterbalance each other.

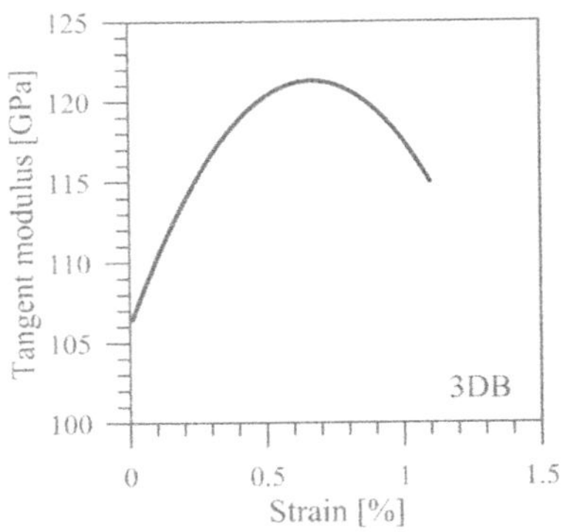

Figure 1.35. *3D braided carbon composite. Quasi-static tensile tests: representative tangent modulus versus strain*

Information on damage initiation and propagation are collected by recording the AE events at each stress/strain level (Figure 1.36) during tension of the 3DB carbon composite. The processing of the energy of the individual AE events provides the curve of cumulative energy, which, as mentioned, presents abrupt slope variations (knee) at the characteristic damage thresholds of the material [LOM 08a]. The damage thresholds (strain ε_1, ε_2 and correspondent stress σ_1, σ_2) from three tests are shown in Figure 1.37. The AE events were registered very early after the start of loading, and damage thresholds are quite lower compared to other textile composites, see, e.g., the glass textile reinforced ones in section 1.3.1 and 1.4.1, and the triaxial braided carbon/epoxy composite in [IVA 09b].

In the strain range of the first AE threshold (ε_1), transverse cracks inside the impregnated yarns and micro fiber–matrix debonding are hypothesized to occur, but direct observations at this strain level have not been performed. At the second strain threshold (ε_2), the development of cracks located at the external yarn boundaries is the dominating damage mechanism, as detected by microscopy observations.

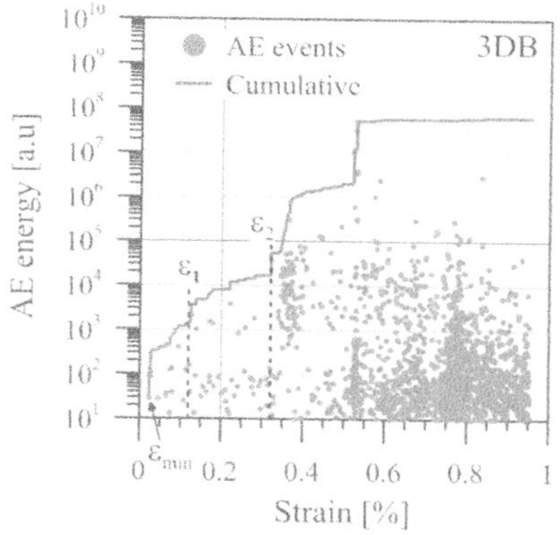

Figure 1.36. *3D braided carbon composite. Quasi-static tensile tests: representative AE energy versus strain diagram*

The damage imparted in three specimens loaded up to the strain level of 0.35% was observed on several cross-sections using a microscope. This strain level was considered to characterize the damage mode in the range of the second AE strain threshold (ε_2). After quasi-static loading (up to the strain of 0.35%) and unloading, the specimen was sliced in several parts with a cut distance of 7 mm, using a diamond saw. Each part was polished and observed on both sides with a microscope.

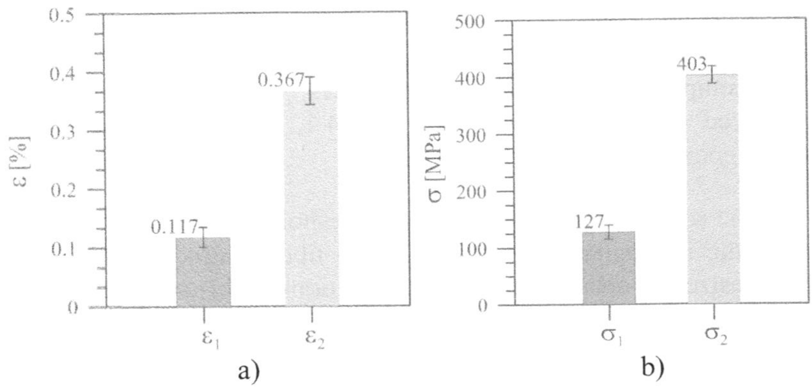

Figure 1.37. *3D braided carbon composite. Quasi-static tensile a) strain and b) stress damage thresholds. Average and standard deviation of three specimens*

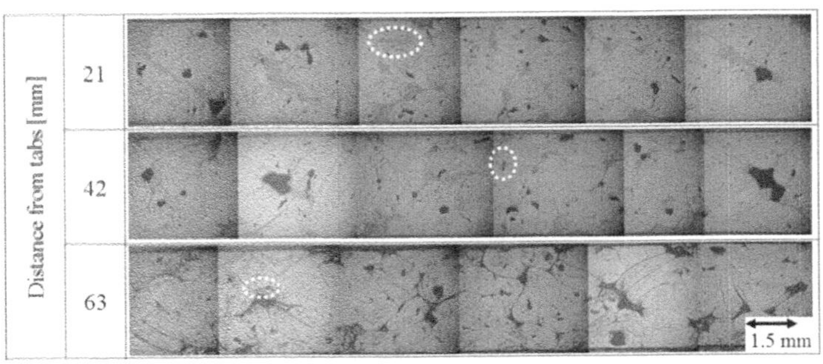

Figure 1.38. *3D braided carbon composite. Quasi-static tensile tests. Damage observation at the strain threshold ε_2: image of cross-sections by optical microscope*

The typical damage for this strain level, in three cross-sections located at three distances from the tabs, is shown in Figure 1.38. The microscopic images show debonding at the interface of yarns and matrix. Some of these cracks at the interfaces are highlighted in the magnifications of Figure 1.39.

Distance from tabs [mm]		
21	42	63

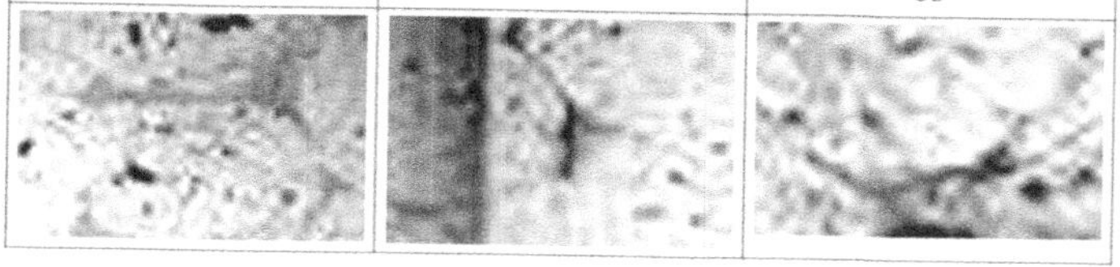

Figure 1.39. *3D braided carbon composite. Quasi-static tensile tests. Magnification of damage in the ellipses in Figure 1.38*

1.5.2. *Fatigue life and damage metrics*

The fatigue tests of the 3DB composite involved different maximum stress (σ_{max}) levels in the cycle, ranging from 950 MPa to the level σ_{inf} for which complete failure of the specimen does not occur after 5 million cycles. As for the other materials in this chapter, the fatigue life diagram (Figure 1.40) has a semilogarithmic representation, with normalized maximum stress to the quasi-static strength (σ_u) for a proper comparison to other materials with different tensile strength.

The load level σ_{inf}, supposed to be the "fatigue limit", was 800 MPa or 60% of the average tensile static stress (see Table 1.4). This is a good performance indicator for the 3DB when compared to the reference level of 65–70% for the unidirectional carbon/epoxy composite detailed in [SIM 03]. The result is more interesting when compared to the level of other carbon textile reinforced composites. The tensile fatigue stress level, which does not impart failure after 2 million cycles, was 40% and 50% of the respective static strength for the non-crimp structurally stitched and the unstitched carbon/epoxy laminates, respectively, described in section 1.6 [CAR 10b]. The value of 53% was measured for the carbon/epoxy non-crimp fabric composite detailed in [VAL 09]. For the non-crimp 3D orthogonal carbon fabric composite, studied in [KAR 11], the value was 50%. Those results indicate that, among the compared composites, the 3DB has the closest σ_{inf} value to the unidirectional composite. As commented in [CAR 10b], this level can be increased for the 3DB reducing the void content and regions where fibers are not fully impregnated.

Comparing the load level σ_{inf} to the AE damage thresholds of the quasi-static loading (see Figure 1.37), it is much higher than the second one σ_2. This could be explained with the different role of the carbon fibers on the fatigue failure of the textile composite, as discussed in Chapter 2.

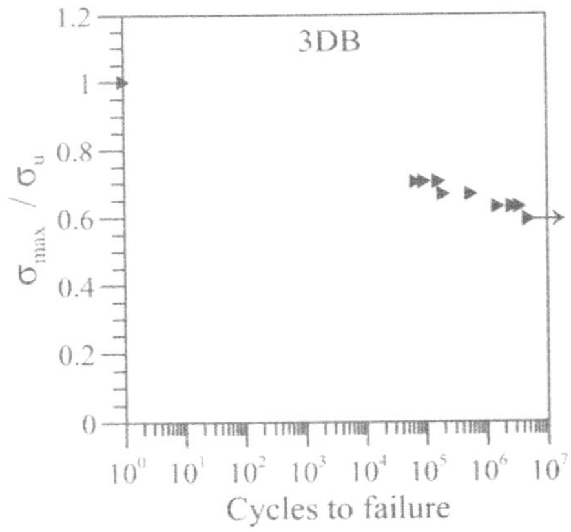

Figure 1.40. *3D braided carbon composite. Fatigue life diagrams: normalized stress (maximum stress in the cycle σ_{max}/quasi-static tensile strength σ_u) versus number of cycles to failure. "→", no failure*

The preliminary understanding of the damage evolution during fatigue loading of the 3DB composites is obtained with the same empirical metrics adopted for the other materials in this chapter. The variation of the shape of the stress displacement cycles during fatigue loading provides the evolution of the "cycle slope" and "cycle dissipation", as defined in section 1.3.2.

For three fatigue load levels, ranging from 60% (σ_{inf} = 800 MPa) to 67% (900 MPa) of the quasi-static tensile strength, the two metrics are represented in Figure 1.41, as a ratio to the respective initial value. The typical initial stage [GAG 06] is visible for the two lowest stress levels at least. The cycle slope has a rapid reduction (growth of the dissipation) showing the fast loss of material stiffness due to the initial damage initiation and evolution in almost the first 10% of the fatigue loading cycles. The second stage of the curves in Figure 1.41, for the two lowest stress levels, has the characteristic slowest degradation of the cycle slope (increase in the dissipation) as a result of the damage evolution up to almost 90% of the fatigue life for the 850 MPa level, and up to the end of the test without failure for σ_{inf}. The third stage of the metric curves covers the remaining part of the fatigue life in which the damage has a fast progression leading to failure (see stress level 850 and 900 MPa in Figure 1.41). The three stages of the fatigue life of the 3DB can be connected to the initiation and development of damage modes, which follow the same evolution observed for the quasi-static loading. The fatigue damage is described in section 1.5.3 with X-ray micro-CT images.

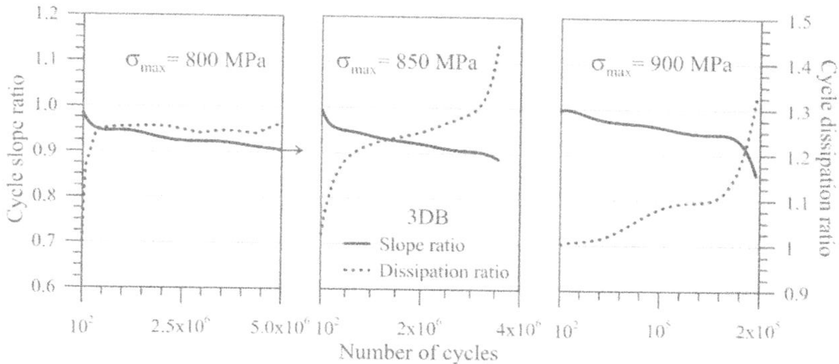

Figure 1.41. *3D braided carbon composite. Comparison of the cycle slope ratio and cycle dissipation ratio for some maximum stress levels. "→", no failure*

1.5.3. *Fatigue damage observation and evolution*

The X-ray micro-CT allowed 3D observations of the damage imparted during tension–tension fatigue tests of 3DB. The damage observation was conducted on specimens cyclically loaded with maximum stress level σ_{inf} (800 MPa). The tests were stopped after 1 million and 3 million cycles for damage examination. The specimens were monitored on three positions (named zone in the pictures) along the free length, achieving a voxel size of 10.3 mm. The damage observations in zone 1 and 2 are displayed in Figure 1.42 and Figure 1.43 in three sections, namely longitudinal (load direction) spanning the width (*x–z*); longitudinal through the thickness (*y–z*); transverse cross-section (orthogonal to the load direction) (*x–y*). After 1 million cycles, several yarn and matrix debondings are visible, which were initiated during the first part of the cyclic loading. As a result of the debondings, the fast reduction of the cycle slope (increase in dissipation) is recorded (Figure 1.41, left). Cracks initiate, within 1 million cycles, both in the resin-rich zones and at the yarn boundaries (Figure 1.42 and Figure 1.43, top images). Increasing the number of cycles up to 3 million, these cracks extend along the matrix–yarn interfaces with splitting of the impregnated yarns. Moreover, other cracks appear and develop that do not initiate from the beginning of the loading (Figure 1.42 and Figure 1.43, bottom images), also visible in the cross-section of zone 3 in Figure 1.44. The elongation of the cracks at yarn–matrix interfaces is related to the slow reduction of the cycle slope (Figure 1.41, left), which means slower and continuous decrease in the material stiffness than that in the beginning of the fatigue loading. The further and new cracks that developed in this late stage of the fatigue loading were responsible for the premature failure of the specimen. In fact, this specimen, considered for damage observations, failed a few cycles prior to the 5 million, even though the applied maximum stress was σ_{inf}. This is probably due to the severe damage state imparted, more critical than the damage in all those specimens, which did not fail under the same maximum cycle stress. A possible explanation is related to the content of voids and dry spots resulting in imperfect impregnation of that 3DB composite bar. As pointed out in [CAR 13], and as a consequence of the latter observations, fatigue life of the 3DB carbon composite could be extended by improving the impregnation technique for the reduction of voids and dry spots in the material.

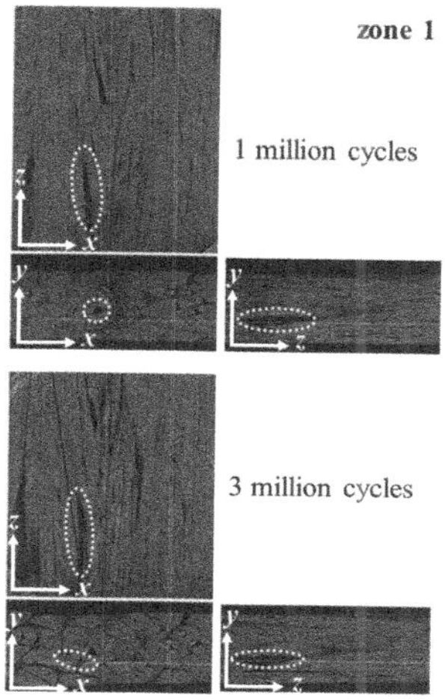

Figure 1.42. *3D braided carbon composite. Fatigue test maximum stress σ_{inf} = 800 MPa. Damage at different cycles: X-ray micro-CT images of zone 1. "z" is the load direction*

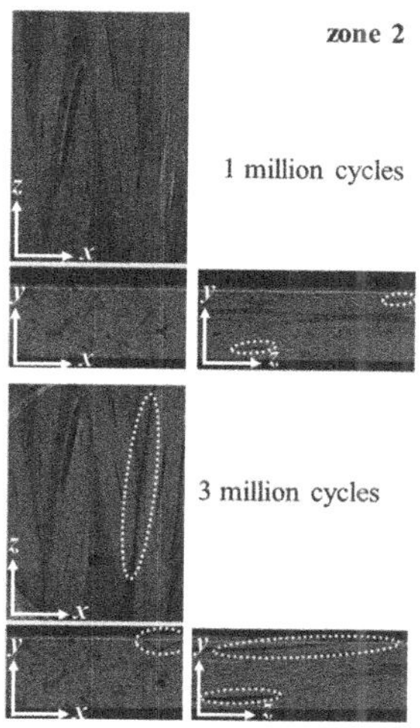

Figure 1.43. *3D braided carbon composite. Fatigue test maximum stress σ_{inf} = 800 MPa. Damage at different cycles: X-ray micro-CT images of zone 2. "z" is the load direction*

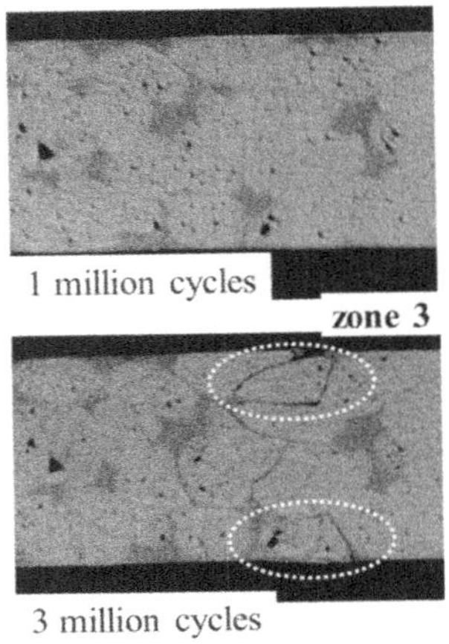

Figure 1.44. *3D braided carbon composite. Fatigue test maximum stress σ_{inf} = 800 MPa. Damage at different cycles: X-ray micro-CT images of a cross-section in zone 3*

1.5.4. *Postfatigue mechanical properties*

The cyclic loading creates diffuse damage as observed in the previous section. The consequences of the fatigue damage on the mechanical properties of the 3DB composite were assessed with postfatigue quasi-static tensile tests after 1, 3 and 5 million cycles with the fatigue stress level of σ_{inf} (800 MPa). The average values, of two specimens for each number of cycles, give a clear trend of the fatigue damage effect on the retention of the main tensile mechanical properties collected in Figure 1.45.

The longitudinal elastic modulus (*E*) decreases continuously with an increasing number of cycles. The reduction is maximum due to the fatigue in the first million cycles (almost 16%). Then, the retention has a slight decrease between 1 and 3 million cycles (2%) and a further reduction of 5% between 3 and 5 million cycles, for a final retention of 77%. This behavior reflects the effect of the damage on the cycle slope as measured during fatigue loading, detailed in Figure 1.41 (left diagram). The damage mode, responsible for the initial reduction of the material stiffness, is the diffuse yarn and matrix deboning described in section 1.5.3. The extension of the deboning along the matrix–yarn interfaces creates the small reduction of the elastic modulus up to 3 million cycles, while the new cracks developed in the late part of the fatigue loading generate the last decrease in *E*.

The reduction of stiffness provides an increase in the deformation at any stress level, clearly visible for the ultimate strain (ε_u) corresponding to the ultimate stress (σ_u) (Figure 1.45).

The damage modes observed during fatigue loading do not show failure of fiber in the yarns for the stress level σ_{inf} after 5 million cycles. This justifies the negligible variation of the quasi-static tensile strength (σ_u) after the three fatigue loading cycles considered (Figure 1.45). The yarns continue to carry the same load level as in the unfatigued material, despite the matrix debonding and internal splitting. The interlacement of the 3DB reinforcement with imparted fatigue damage still ensures the proper internal load transfer.

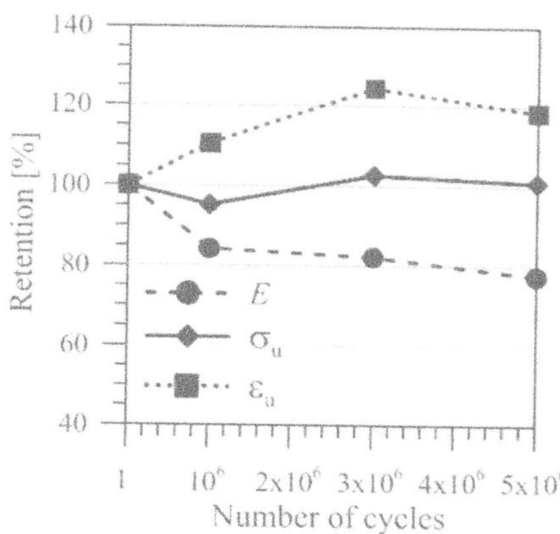

Figure 1.45. *3D braided carbon composite. Postfatigue quasi-static tensile tests. Retention of the elastic modulus (E), strength (σ_u) and ultimate strain (ε_u)*

The similarities and differences of the damage evolution during pre- and postfatigue quasi-static tensile tests are perceived with the registration of AE events. The first difference is the larger amount of AE events recorded since the early stage of the postfatigue test

(Figure 1.46 left) compared to the prefatigue one (Figure 1.36). This is probably due to the further continuous increase in the crack length at the yarn/matrix interfaces, previously initiated during the 5 million fatigue cycles at the load level of σ_{inf}. In fact, during quasi-static tensile test below the load stress of σ_{inf}, new damage is not supposed to be created because the damage up to that load level was anticipated in the fatigue loading. The huge quantity of AE events, in the late state of the quasi-static test above σ_{inf}, can be consequence of new activated damage mechanisms leading, for increasing load, to the ultimate stress, which is practically independent of the number of preliminary fatigue loading cycles. Differently from the postfatigue behavior of the glass fiber textile composites in this chapter (sections 1.3.4 and 1.4.4), the cumulative AE energy curve of the postfatigue quasi-static loading of the 3DB carbon composite is shifted slightly to a higher stress/strain level of the AE initiation. Moreover, an interesting result is the translation of the cumulative curve to higher energy levels. In postfatigue loading, the 3DB maintains similar damage thresholds but with higher energy levels. This can be explained with the activation of high energy damage mechanisms with lower load levels than the prefatigue that, however, do not mean a premature failure of the material.

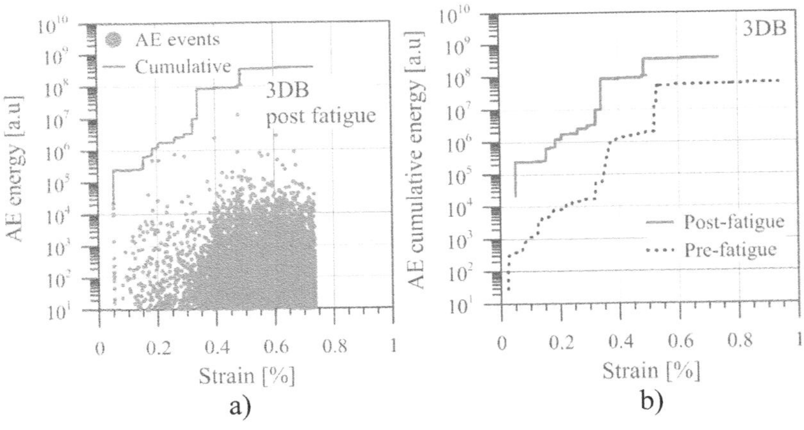

Figure 1.46. *3D braided carbon composite. Postfatigue quasi-static tensile test after 5 million cycles. a) Representative AE energy versus strain diagram. b) Comparison of the pre- and postfatigue cumulative energy*

1.6. Fatigue behavior and damage evolution in non-crimp stitched and unstitched carbon reinforced epoxy composite

The stitched and unstitched preforms are made up of two Seartex® NCFs with Tenax HTS carbon fibers. The fabrics have plies construction +45°/–45° (total areal density of 540 g/m^2) and 0°/90° (total areal density of 556 g/m^2). A polyester sewing thread (8.3 tex, 2.6 × 5 mm tricot + chain knitting pattern, 6 g/m^2) knits together the carbon plies. The non-structural sewing is in the direction 90° (see Figure 1.47). The adopted stacking sequence is symmetric [45/–45/0/90/45/–45]$_S$ with a total thickness of about 4.2 mm before impregnation.

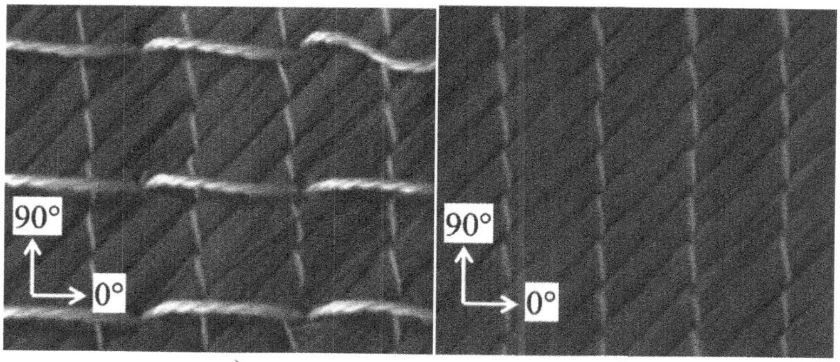

a) b)

Figure 1.47. *Non-crimp a) structural stitched and b) unstitched carbon preform*

The structurally stitched preform has a 5 × 5 mm square piercing pattern made with 1 K carbon rowing and tufting method (KL RS 522 stitching head mounted on a KUKA-robot). The structural stitching direction corresponds to 0° (Figure 1.47(a)). Here, "structural" means that the stitching yarn does not only consolidate the plies (as the non-structural one) but also forms through the thickness reinforcement. As it is well known and studied (see for similar preform [LOM 02]), the process of stitching results in local perturbation of the uniform placement of the fibers, causing resin-rich regions in the composite and affecting the mechanical performance of the composite. The stitching-induced openings in the preform are naturally oriented along the global fibers orientation in the ply, as observed and measured in [KOI 09]. Some features of the carbon yarns in the fabrics, the stitching yarn and the non-structural sewing thread are collected in Table 1.5.

A vacuum-assisted process was adopted for the impregnation of the preforms using RTM-6 epoxy resin (HexFlow®–Hexcel®). The thickness of the cured laminates was 3.2–3.5 mm. This shows a significant compaction with respect to the dry state. The measured fiber volume fraction was 54% [CAR 10b, KOI 09].

The tensile quasi-static and fatigue behavior of the carbon structurally stitched and unstitched composites involved the experimental setup described in section 1.2, including AE and DIC. The fatigue damaged specimens, after a predefined number of cycles, were inspected with the same X-ray micro-CT system mentioned in section 1.5 for the 3D braided carbon composite.

In the following, for brevity, the composite tufted with the carbon structural yarn and the polyester thread is named "stitched", while the composite with only the polyester thread is name "unstitched".

	Tenax® HTS	Tenax® HTS (stitching yarn)	Polyester
Linear density (tex)	800	67	8.3
Number of fibers	12,000	1,000	12
Twisting (m^{-1})	0	S15	Z24

Table 1.5. *Features of the yarns in the fabrics, the stitching yarn and the non-structural sewing thread*

1.6.1. *Quasi-static tensile behavior*

The quasi-static tensile behavior, here summarized, was investigated in [KOI 09] and [CAR 09]. For an in-depth discussion of the quasi-static performance and damage observation, the reader is referred to [KOI 09] in which an extensive investigation on the effect of the stitching density is detailed. The main tensile mechanical properties of the two NCF carbon composites for the two loading directions 0° and 90°, shown in Figure 1.48, are: the Young's modulus (*E*), ultimate stress (σ_u) and ultimate strain (ε_u). In both directions, the unstitched laminate shows a slight higher average stiffness, while the 90° direction has elastic modulus higher than the 0° one for both composites (Figure 1.48, left). However, the structural stitching has a minor effect on the in-plane stiffness, with values almost in the same scatter band.

The structural stitching is really effective when the ultimate properties are considered. It improves the tensile strength for both load directions compared to the unstitched counterpart (Figure 1.48, center). The contribution of the structural stitching in carrying the load is visible as the highest strength in the stitching direction (i.e. 0°). The improvement of the ultimate tensile properties also includes the ultimate strain (Figure 1.48, right). This stitched carbon NCF composite can be considered one of the exceptions in the wide range of stitched composites considered in [MOU 08] and [MOU 10]. In fact, the stitching improves the ultimate properties of the material. As argued in [KOI 09], the increase in strength should be attributed to the stitching, which enhances the interlaminar fracture toughness. The stitching increases the through-the-thickness strength and delays local delamination. This is a relevant aspect with some effects on the fatigue performance detailed in section 1.6.2.

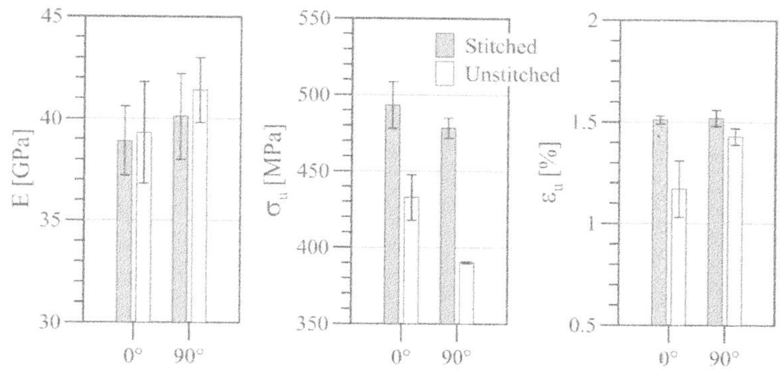

Figure 1.48. *Stitched and unstitched NCF carbon composite. Quasi-static tensile properties from [KOI 09]*

The influence of the stitching on the damage initiation and development was assessed with the AE events. The recording during quasi-static loading provides the AE cumulative energy curves compared in Figure 1.49, for both NCF composites and load directions. Low energy events start to occur at a lower strain level for the stitched than the unstitched laminate for both directions of the applied load. It means an early onset of the damage in the tufted material. The tendency is confirmed with the increase in the load. Extracting from the AE cumulative energy curves the stress (strain) levels at the abrupt slope variations (i.e. "knees"), the two characteristic damage thresholds show higher levels for the unstitched composite (Figure 1.50). The damage modes attributed to the two thresholds in [KOI 09] are as follows: for ε_1, growth of narrow cracks and initiation of new cracks in the weakest locations; for ε_2, extensive appearance of relatively large cracks, presumably in the off-axis plies. The structural stitched composite has an early development of those damage modes at the beginning of the loading process. At the late stage of the loading, the trend change, the AE cumulative energy of the unstitched material continue to increase, while remaining almost stable for the stitched one. This confirms a continuous and fast development of the damage in the unstitched composite and a slower evolution in the stitched one, leading to the different strengths in Figure 1.48.

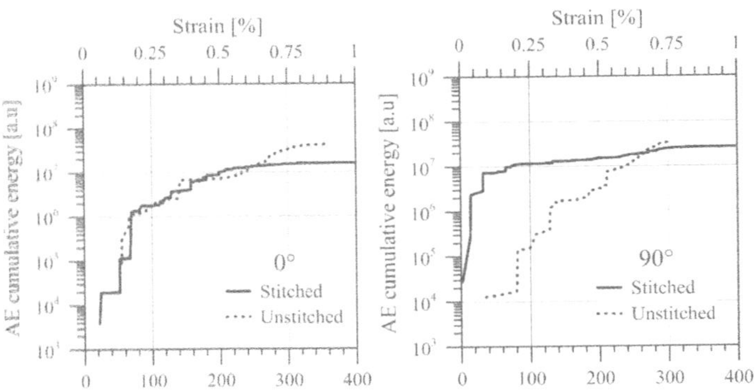

a) b)

Figure 1.49. *Stitched and unstitched NCF carbon composite. Quasi-static tensile tests: representative AE cumulative energy versus stress/strain diagrams for a) 0° direction; b) 90° direction; from [KOI 09]*

The X-ray images in [KOI 09], for different strain levels, exhibit crack initiation at the polyester sewing thread insertion sites of the non-structural stitched composite. The carbon structural stitching creates larger clusters of cracks mainly in the plies oriented in ±45° to the loading direction and few cracks in 0°/90° plies. This suggests that the larger width of structural stitching openings in ±45° plies has a dominant influence on crack initiation and growth.

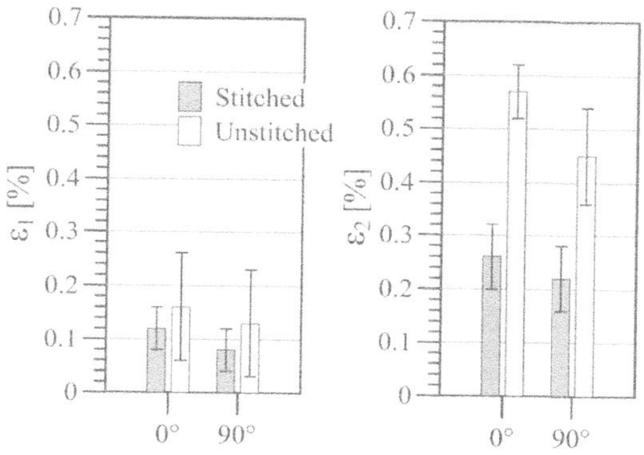

Figure 1.50. *Stitched and unstitched NCF carbon composite. Quasi-static tensile strain damage thresholds from [KOI 09]*

1.6.2. *Fatigue life and damage metrics*

The fatigue life diagram (maximum stress in the cycle σ_{max} vs. number of cycles to failure) of both stitched and unstitched NCF carbon composites includes several stress levels from the range of low cycles fatigue to the level σ_{inf} (named "fatigue limit" in the context discussed in Chapter 2). For the two composites considered, the stress level σ_{inf} was assumed for the cyclic loading without specimen failure after 2 million cycles. The fatigue life diagrams, for the 0° (Figure 1.51(a)) and 90° (Figure 1.51(b)) loading, include the stress level σ_{inf} (200 MPa), which is the same for both materials and load directions.

It is worth mentioning that the load level σ_{inf} of the stitched laminates is higher than the second AE damage threshold σ_2 (ε_2) of the quasi-static loading (see Figure 1.48), while it is comparable for the unstitched composite. This is commented on in Chapter 2 which considers the distinct role of the carbon fibers on the fatigue failure of the textile composites.

The different fatigue behavior of the stitched and unstitched composites is clearly visible as shown in Figure 1.51, mainly for the 90° loading direction. A direct comparison of the average fatigue life highlights the differences not only for the 90° but also for the 0° load condition. The carbon stitching yarns aligned in the 0° direction contributes to extending the fatigue life. For any stress level considered, the consequence of the stitching yarn in carrying the load is a longer fatigue life of the composite loaded in 0° direction than in 90° (Figure 1.52, top). In contrast, the stitching yarn contributes to shorten the fatigue life when load is applied in 90° direction. This is connected to the wider opening introduced during stitching that are resin-rich zones in the composites, as measured in [KOI 09].

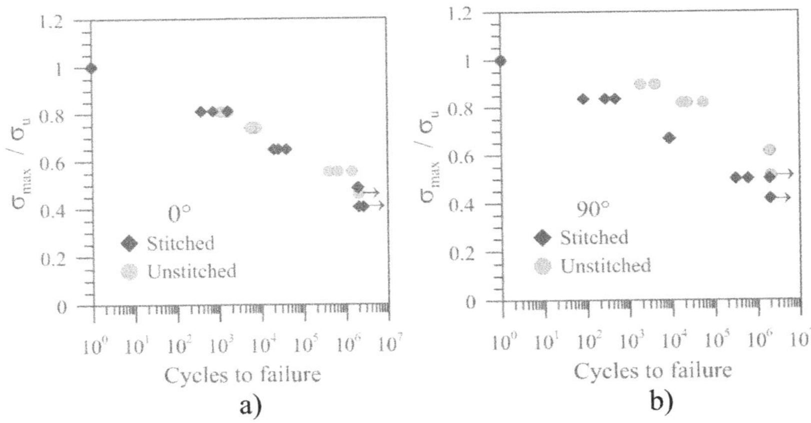

Figure 1.51. *Stitched and unstitched NCF carbon composite. Fatigue life diagrams: normalized stress (maximum stress in the cycle σ_{max}/quasi-static tensile strength σ_u) versus number of cycles to failure. a) 0° and (b) 90° direction loading. "→", no failure*

Contrary to that, the unstitched composite has a longer fatigue life when loaded in the 90° than in the 0° direction, for any considered stress level (Figure 1.52, bottom). This shows that the best performance in term of fatigue life is in the direction of the polyester sewing (90°, see Figure 1.47), and the imparted openings have a predominant effect on reducing the fatigue life in the direction orthogonal to their length. As for the quasi-static loading, the structural stitching overcomes this drawback and, moreover, extends the fatigue life of the laminate in the direction of the tufted yarn. Conversely, the structural stitching reduces the fatigue life when cyclically loaded orthogonal to its direction, while the quasi-static strength is improved (Figure 1.48), as well.

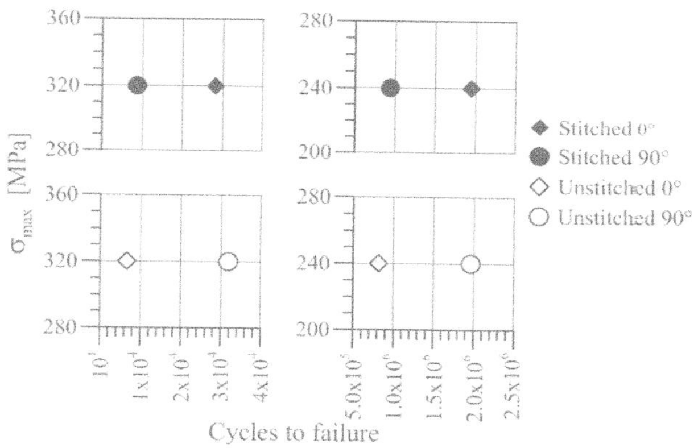

Figure 1.52. *Stitched and unstitched NCF carbon composite. Comparison of the average fatigue life for some maximum stress levels*

The measurements of the two NCF composites during cyclic loading make it possible to evaluate the damage metrics, adopted through this chapter, in order to have an initial understanding on the different evolution of the fatigue damage. The "cycle slope" and "cycle dissipation", defined in section 1.3.2, are compared for both materials and directions for the fatigue load level of σ_{inf}. The different behavior observed with the fatigue life diagrams is here underlined with the different damage development, as the cycle slope qualitatively shows in Figure 1.53 and the cycle dissipation in Figure 1.54 (ratio to the respective initial value). Faster evolution of the damage in the stitched NCF composite is observed in the initial part of the cyclic loading, almost 10% of the fatigue cycles in 0° direction. This results in a fast reduction of the cycle slope (Figure 1.53, left). Analogously, the damage in the unstitched composite starts to have a rapid diffusion from the beginning of the 0° loading, but, differently, the same fast evolution continues up to almost 40% of the fatigue loading cycles, with constant rate of decrease in the cycle slope (Figure 1.53) and increase in the dissipation (Figure 1.54). The damage mode, initiated in this first part of the fatigue loading, creates a reduction of the material stiffness (comparable to the cycle slope) of 15% and 35% in the stitched and unstitched NCF composite, respectively (Figure 1.53, left). This observation gives a preliminary understanding on the effectiveness of the stitching to localize and constrain the diffusion of the damage. For the reaming part of the relative 0° fatigue loading, the cycle slope has similar continuous reduction (increase in dissipation) with a much slower rate, showing different diffusion mechanism of the damage in both composites. The damage imparted in this second stage of 0° fatigue loading generates further almost 5% reduction of the stitched and unstitched NCF composite stiffness (see cycle slope in Figure 1.53 left), respectively. The latter contribution to the stiffness degradation points out the modification of the damage modes during fatigue life, as presented in section 1.6.3 by X-ray pictures, and their different impact on the deterioration of some mechanical properties.

The qualitative distinction of the two stages damage evolution is also detectable for the 90° fatigue loading. As for the fatigue life diagram, the completely different behavior is for the stitched composite in the 90° direction. The cycle slope (dissipation) shows a rapid initiation of the damage in almost 10% of the fatigue loading cycles, then the degradation (grow of dissipation) continues similarly up to 1.5 million cycles. The cyclic loading in 90° direction of the unstitched material produces the largest reduction (increment) of the cycle slope (dissipation) in the first quarter of the total loading cycles, while in the remaining part a different damage diffusion is supposed with the lower variation rate of the cycle slope (Figure 1.53, right) and dissipation (Figure 1.54, right). The comparison of the two behaviors for 90° loading suggests different propagation of the fatigue damage modes, which finally produce the same stiffness (cycle slope) loss (≈55%) of the stitched and unstitched NCF composites.

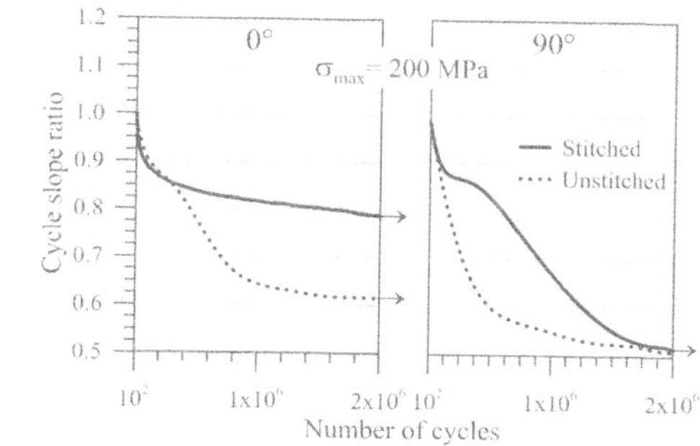

Figure 1.53. *Stitched and unstitched NCF carbon composite. Comparison of the cycle slope ratio for fatigue stress of σ_{inf}. "→", no failure*

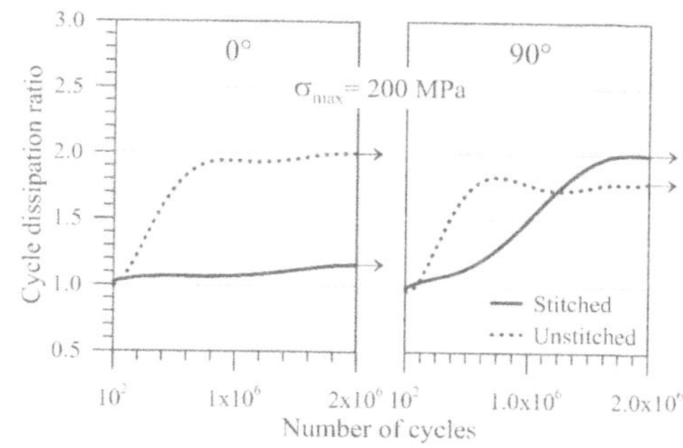

Figure 1.54. *Stitched and unstitched NCF carbon composite. Comparison of the cycle dissipation ratio for fatigue stress of σ_{inf}. "→", no failure*

1.6.3. *Fatigue damage observation and evolution*

The fatigue damage is here described for loading in 0° direction with a maximum stress level of 240 MPa. The X-ray damage examination was after three levels of fatigue cycles (1,000, 10,000 and 100,000) in a portion located in the center of the specimens covering the complete width. The three levels were selected to be within the initial 10% of the fatigue life, where the damage metric diagrams show the main concentration of the damage development. It is worth mentioning that the average fatigue life, for this load level and direction, of the stitched composite is more than double of the unstitched one (see Figure 1.52). Moreover, for the considered load direction 0°, the stitched material has the best while the unstitched has the worse fatigue performance compared to the 90° loading.

The front and side X-ray images in Figure 1.55 show, after 1,000 cycles, cracks in ±45° directions mostly affecting the external layers of the stitched and unstitched NCF composites (see first row of Figure 1.55). Density of those cracks is larger in the stitched material. With the increase in fatigue cycles, the density of cracks rises and after 10,000 cycles some openings appear on the stitched composite located at the insertion positions of the tufted yarn, according to the stitching pattern (see second row of Figure 1.55). The diffusion of the ±45° cracks and the latter openings are mostly responsible for the postfatigue strength degradation of the material, while the stiffness seems to be unaffected, as shown in section 1.6.4. After 10,000 cycles, the unstitched composite exhibits diffuse pattern of the ±45° cracks with growing length.

The detected damage mode in the unstitched composite, after 100,000 cycles, is large delamination (see bottom row of Figure 1.55). The delamination zones are mainly diffused at the interface between the ±45° external layers, and debondings are also visible in the inner part of the thickness. This evolution of the damage mode contributes to the stiffness and strength degradation in the remaining part of the fatigue life leading to complete failure. After the same number of cycles, the stitched composite exhibits different behavior. The damage visible in the thickness confirms that the stitching reduces the delamination. Some delaminations appear after 10,000 cycles, but are extremely reduced in comparison to the unstitched composite (see bottom row of Figure 1.55).

The positive effect of the structural stitching on the imparted fatigue damage is also clear when looking at the thickness surface of specimens fatigued for 2 million cycles with the load level of σ_{inf}. Figure 1.56 shows the extremely diffused delamination on the unstitched composite for both load directions with large zones of debonding, while the stitched counterpart, 90° loaded, has limited delamination and the laminate is still compact. The corresponding picture of the 0° loaded stitched NCF composite is not displayed, for which delamination is less visible than the 90° loading case.

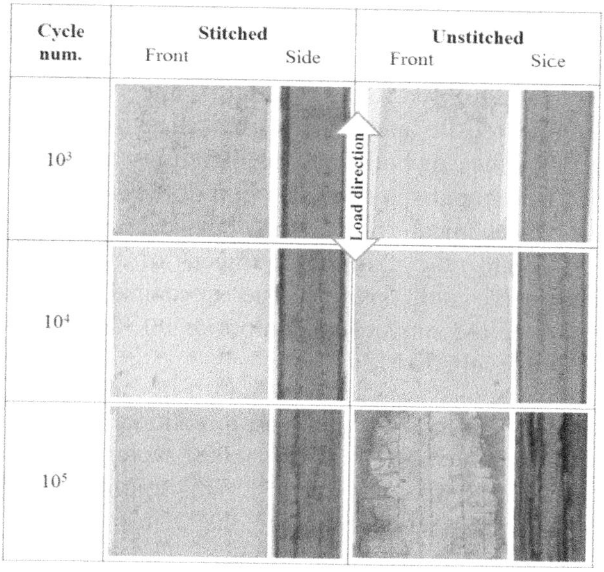

Figure 1.55. *Stitched and unstitched NCF carbon composite. Fatigue test maximum stress 240 MPa in 0° direction. Damage at different cycles: X-ray micro-CT images of a portion in the specimen center*

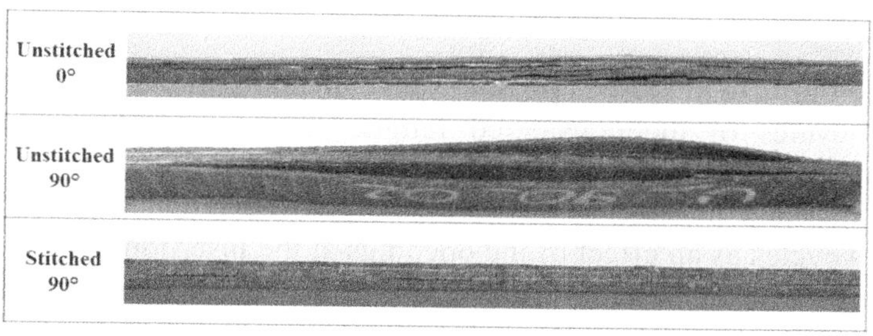

Figure 1.56. *Stitched and unstitched NCF carbon composite. Fatigue test maximum stress 200 MPa after 2 million cycles. Thickness view of representative specimens*

1.6.4. *Postfatigue mechanical properties*

The specimens adopted for the damage observations in the previous section were considered to assess the influence of the fatigue loading on the tensile mechanical properties. Postfatigue quasi-static tensile tests provide this information using specimens fatigued with the stress level of 240 MPa for 1,000, 10,000 and 100,000 cycles. The retentions of the tensile mechanical properties of the stitched and unstitched composites in both loading directions are summarized in Figure 1.57. The comparison shows a trend of the fatigue influence on the residual mechanical properties, but cannot be considered exhaustive due to the limited number of specimens. Some measurements were not possible due to damage imparted after 100,000 cycles of the unstitched composite 90° loaded (see Figure 1.56 for the load level 200 MPa).

Irrespective of the material and load direction, the elastic modulus does not have relevant variation up to 10,000 cycles (variations can be considered in the experimental scatter). This implies that the fatigue imparted ±45° cracks, observed by the X-ray pictures, have a negligible effect on the residual stiffness. The observed extensive delamination in the 0° loaded unstitched composite, after 100,000 cycles (Figure 1.55), has as a consequence a relevant reduction of the stiffens, almost 40%, while the same number of cycles imparts localized and reduced delamination in the stitched composite (Figure 1.55), which does not considerably affect its stiffness.

The ultimate properties do not change meaningfully after the initial 1,000 cycles of the considered fatigue stress level (Figure 1.57) (variations can be considered in the experimental scatter). The stitched material starts to have notable reduction of the ultimate stress after 10,000 cycles as an effect of the openings at the insertion of the tufted yarn, which continue to decrease up to 100,000 cycles as a consequence of the onset of localized delamination. The diffuse delamination in the unstitched material (Figure 1.55) generates almost 40% decrease in the 0° direction strength, while the orthogonal direction, showing better fatigue performance, has half of that reduction. The postfatigue tensile strength highlights the effectiveness of the structural stitching in suppressing the delamination with a loss of ultimate stress less than half of the unstitched counterpart.

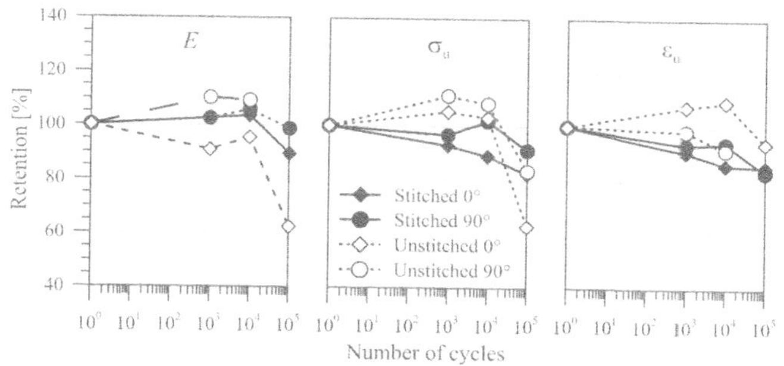

Figure 1.57. *Stitched and unstitched NCF carbon composite. Postfatigue quasi-static tensile tests. Retention of the elastic modulus (E), strength (σ_u) and ultimate strain (ε_u)*

The evolution of the damage imparted during the postfatigue quasi-static tensile test was considered with the registration of AE events and the DIC strain mapping. For a load level (220 MPa) lower than the one adopted in the fatigue test (240 MPa), the distributions of the strain component in the load direction for both materials and load direction are quite similar to those of the prefatigue loading presented in [KOI 09]. The strain component maps of specimens fatigued for 10,000 cycles show a regular pattern on the external surface of the stitched composite (Figure 1.58, left) with concentrations in the insertion positions of the carbon yarn, where openings were detected after the fatigue loading. The strain maps of the unstitched composite are almost uniform: the polyester sewing thread does not get appreciable peaks with the adopted images resolution.

Typical AE diagrams for postfatigue quasi-statically loaded specimens (after 10,000 cycles) are depicted in Figure 1.59 and Figure 1.60 for 0° loaded stitched and unstitched composite, respectively. The diagrams show a significant amount of events recorded close to the maximum stress formerly set during fatigue tests (240 MPa), below this threshold the amount of noise is barely significant. The damage generated during prefatigue quasi-static tensile tests, below the stress level of 240 MPa, was not detected in these AE recordings because it was completely imparted during cyclic tests. Further damage is not generated in the initial stage of the quasi-static loading. This is linked to the negligible variation of the postfatigue elastic modulus (Figure

1.57), which is mainly due to the fatigue damage observed as ±45° cracks on the external layers.

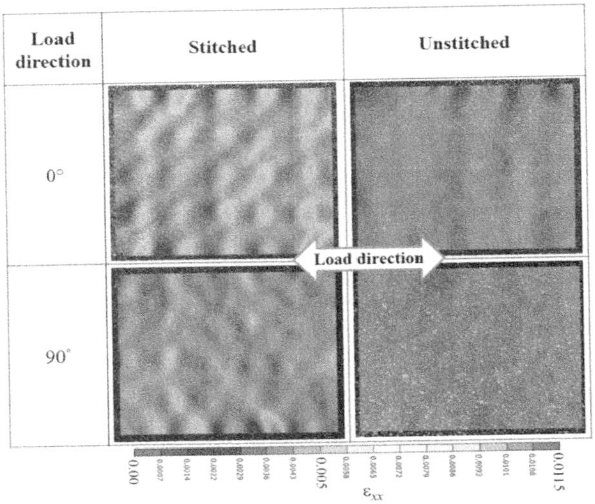

Figure 1.58. *Stitched and unstitched NCF carbon composite. Postfatigue quasi-static tensile tests. Map of the strain component in the load direction (ε_{xx}) for a load level of ≈220 MPa after 10,000 cycles (vertical side covers the entire width)*

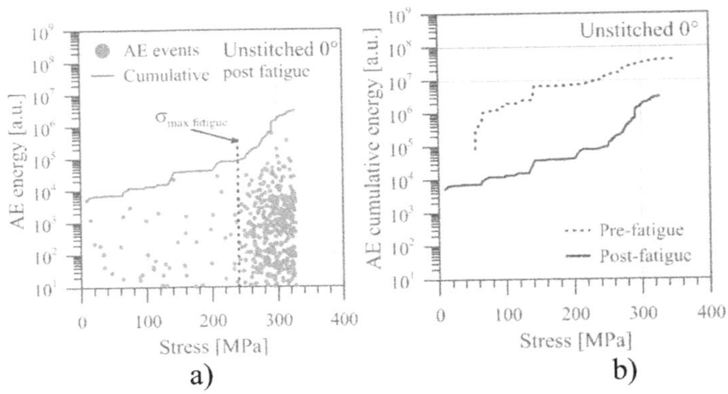

Figure 1.59. *Unstitched NCF carbon composite. Postfatigue quasi-static tensile test after 10,000 cycles, maximum stress 240 MPa, 0° direction. a) Representative AE energy versus stress diagram. b) Comparison of the pre- and postfatigue cumulative energy*

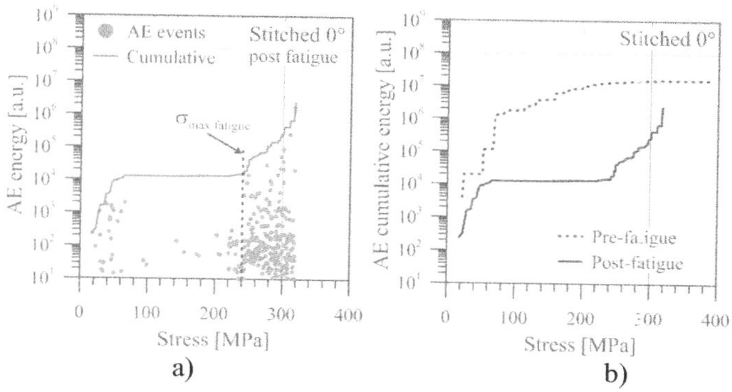

Figure 1.60. *Stitched NCF carbon composite. Postfatigue quasi-static tensile test after 10,000 cycles, maximum stress 240 MPa, 0° direction. a) Representative AE energy versus stress diagram. b) Comparison of the pre- and postfatigue cumulative energy*

The number and energy of the AE events increase rapidly above the fatigue load level, when the damage starts to develop. This provides a steep slope increment of the cumulative AE energy curve (Figure 1.59(a) and Figure 1.60(a)). The comparison of the prefatigue [KOI 09] and postfatigue cumulative AE energy curves shows some differences between the two materials loaded in the 0° direction. The unstitched composite had an early initiation of the damage, with AE events of lower energy, compared to the prefatigue counterpart (Figure 1.59(b)). This is probably due to the activation of matrix cracks not completely developed during fatigue loading. The stitched composite shows almost the same load level for the initiation of the AE recording, as in the prefatigue test, with lower energy events (Figure 1.60(b)). Those can be connected to the onset of new matrix cracks in the fatigue damaged configuration of the materials containing diffuse ±45° cracks, and onset of delamination. In the full range of quasi-static loading, up to the removal of the sensors, the cumulative curves of both materials grow with lower energy levels than the prefatigue ones (Figure 1.59 and Figure 1.60). This shows that the damage modes developed, up to the fatigue load, are still typical of low and medium energy levels, namely cracks inside the yarns or on the yarn boundaries and transverse cracks in resin rich pockets, both evolving in local delamination. These are the initial stages of the damage development in textile composites deeply discussed in the Chapter 2.

1.7. Remarks and perspectives

The chapter gives an overview of the tension–tension (R = 0.1) fatigue behavior of a wide range, not exhaustive, of textile composites: from 2D to 3D glass textile reinforced epoxy, 3D braided carbon composite, non-crimp fabric stitched and unstitched carbon laminates.

The experimental data collected here show that an in-depth understanding of the fatigue performance and the fatigue damage evolution needs an extended knowledge including three steps. The first step considers the prefatigue quasi-static tensile behavior of the material, which also provides the load levels (thresholds) of initiation and development of damage modes for a preliminary understanding on the fatigue damage growing. The second step covers the real fatigue loading, in a wide range of load levels, aiming at the fatigue life curve and, moreover, to the observation of the damage development with possible similarities to the quasi-static one. Finally, in the third step, the postfatigue quasi-static loading of differently fatigued specimens provides an overview on the effect of the fatigue on the mechanical properties and on the modification of the initiation and development of the damage mechanisms.

This chapter highlights the influence of the reinforcement architecture on the fatigue performance of textile composites and on the evolution of damage during cyclic loading. The latter can be preliminary comprehended with a correct understanding of the damage modes sequence for increasing quasi-static loading (see the damage development stages description in Chapter 2). This gives an overview of the fatigue damage evolution in term of cycle number. The driving force of the damage development under quasi-static loading is the increasing load corresponding to increasing deformation energy, while under fatigue loading the damage evolves with the growing deformation energy of the loading cycles.

As for the materials in this chapter, the quasi-static damage stress/strain thresholds, assessed by AE events recording, could provide an initial estimation of the "fatigue limit", namely the fatigue load level for which the imparted damage does not lead to failure for a given number of cycles. As discussed in detail in Chapter 2, the first quasi-static damage threshold, connected to the transverse cracks inside the yarns or on the yarn boundaries, is for some textile composites, see the glass reinforced in this chapter, very close to the "fatigue limit". But this correlation cannot be extended to any textile composite. In fact, the carbon-reinforced composites in this chapter have a fatigue limit higher than the first quasi-static damage threshold. The second damage threshold (related to the local delamination and development of long transverse cracks) can be better considered close to their "fatigue limit". This is related to the intrinsic nature of the components, as commented in Chapter 2, which links the fatigue life of the textile composite more to the fatigue of the carbon fibers than the damage of the matrix.

Therefore, as consequence of the results and discussions in this chapter and in Chapter 2, the perspectives to better understand the complex fatigue behavior of textile composites are, in the authors' opinion, mainly included in two investigation lines covering experimental and predictive techniques.

Accurate non-destructive experimental techniques are worthwhile to precisely determine the quasi-static damage load levels. The correlation of the damage threshold and the related damage mode allows for a better understanding of the initial damage and of the fatigue load level, considered as the "fatigue limit" of the textile composite. In this context, recently, the AE events cluster analysis technique has shown an accurate identification of the damage modes in textile composites (see, e.g., [LI 14, LI 15, LI 16]).

The above-mentioned experimental techniques are supposed for synergistic investigations including predictive models of the fatigue damage initiation and development in textile composites. Analytical and numerical predictive tools are indispensable due to time-consuming experimental fatigue testing. Models dedicated to the fatigue of composites and composite structures [DEG 01, VAS 10] and, in particular, textile composites, are continuously under development [LOM 15, VAN 02, VAN 05, VAN 15]. The complexity of the fatigue damage evolution in textile composites requires a continuous interaction of the experimental measurements and observations and the multiscale models to get a reliable prediction of the fatigue life and fatigue damage in such materials.

1.8. Bibliography

[ANO 15] ANOSHKIN A.N., ZUIKO V.YU., "Experimental methods and standards for fatigue of fiber-reinforced composites", in CARVELLI V., LOMOV S.V. (eds), *Fatigue of Textile Composites*, Woodhead Publishing Limited, Cambridge, 2015.

[BOG 09] BOGDANOVICH A.E., MOHAMED M.H., "Three-dimensional reinforcement for composites", *SAMPE Journal*, vol. 45, pp. 8–28, 2009.

[CAR 09] CARVELLI V., KOISSIN V., KUSTERMANS J. *et al.*, "Progressive damage in stitched composites: Static tensile tests and tension-tension fatigue", *17th International Conference on Composite Materials ICCM-17*, Edinburgh, UK, July 27–31, 2009.

[CAR 10a] CARVELLI V., GRAMELLINI G., LOMOV S.V., *et al.*, "Fatigue behavior of non-crimp 3D orthogonal weave and multi-layer plain weave E-glass reinforced composites", *Composites Science and Technology*, vol. 70, pp. 2068–2076, 2010.

[CAR 10b] CARVELLI V., NERI TOMASELLI V., LOMOV S.V. *et al.*, "Fatigue and post-fatigue tensile behaviour of non-crimp stitched and unstitched carbon/epoxy composites", *Composites Science and Technology*, vol. 70, pp. 2216–2224, 2010.

[CAR 13] CARVELLI V., PAZMINO J., LOMOV S.V. *et al.*, "Quasi-static and fatigue tensile behavior of a 3D rotary braided carbon/epoxy composite", *Journal of Composite Materials*, vol. 47, pp. 3195–3209, 2013.

[CAR 16] CARVELLI V., BETTI A., FUJII T., "Fatigue and Izod impact performance of carbon plain weave textile reinforced epoxy modified with cellulose microfibrils and rubber nanoparticles", *Composites Part A*, vol. 84, pp. 26–35, 2016.

[CUR 68] CURTIS G.J., MILNE J.M., REYNOLDS W.N., "Non-Hookean behaviour of strong carbon fibres", *Nature*, vol. 220, pp. 1024–1025, 1968.

[DEB 08] De Baere I., Van Paepegem W., Degrieck J., "Design of mechanical clamps with extra long wedge grips for static and fatigue testing of composite materials in tension and compression", *Experimental Techniques*, vol. 32, pp. 62–69, 2008.

[DEB 09] De Baere I., Van Paepegem W., Degrieck J., "On the design of end tabs for quasi-static and fatigue testing of fibre-reinforced composites", *Polymer Composites*, vol. 30, pp. 381–390, 2009.

[DEG 01] Degrieck J., Van Paepegem, W., "Fatigue damage modelling of fibre-reinforced composite materials: review", *Applied Mechanics Reviews*, vol. 54, pp. 279–300, 2001.

[GAG 06] Gagel A., Fiedler B., Schulte K., "On modelling the mechanical degradation of fatigue loaded glass-fibre non-crimp fabric reinforced epoxy laminates", *Composites Science and Technology*, vol. 66, pp. 657–664, 2006.

[GAR 16] Garcea S.C., Sinclair I., Spearing S.M., "Fibre failure assessment in carbon fibre reinforced polymers under fatigue loading by synchrotron X-ray computed tomography", *Composites Science and Technology*, vol. 133, pp. 157–164, 2016.

[GUD 10] Gude M., Hufenbach W., Koch I., "Damage evolution of novel 3D textile reinforced composites under fatigue loading conditions", *Composites Science and Technology*, vol. 70, pp. 186–192, 2010.

[HAR 03] Harris B., "A historical review of the fatigue behaviour of fibre-reinforced plastics", in Harris B. (ed.), *Fatigue in Composites*, Woodhead Publishing Limited, Cambridge, 2003.

[KAR 11] Karahan M., Lomov S.V., Bogdanovich A.E. *et al.*, "Fatigue tensile behavior of carbon/epoxy composite reinforced with non-crimp 3D orthogonal woven fabric", *Composites Science and Technology*, vol. 71, pp. 1961–1972, 2011.

[KAW 15] Kawai M., "2D woven fabric composites under fatigue loading of different types and in different environmental conditions", in Carvelli V., Lomov S.V. (eds), *Fatigue of Textile Composites*, Woodhead Publishing Limited, Cambridge, 2015.

[KOI 09] Koissin V., Kustermans J., Lomov S.V. *et al.*, "Structurally stitched NCF preforms: quasi-static response", *Composites Science and Technology*, vol. 69, pp. 2701–2710, 2009.

[INO 00] INOUE A., FUJII T., "Effect of loading path on mechanical response of a glass fabric composite at low cyclic fatigue under tension/torsion biaxial loading", *Journal of Reinforced Plastics and Composites*, vol. 19, pp. 111–123, 2000.

[IVA 09a] IVANOV D.S., LOMOV S.V., BOGDANOVICH A.E. *et al.*, "A comparative study of tensile properties of non-crimp 3D orthogonal weave and multi-layer plain weave E-glass composites. Part 2: Comprehensive experimental results", *Composites: Part A*, vol. 40, pp. 1144–1157, 2009.

[IVA 09b] IVANOV D.S., BAUDRY F., VAN DEN BROUCKE B. *et al.*, "Failure analysis of triaxial braided composite", *Composites Science and Technology*, vol. 69, pp. 1372–1380, 2009.

[LEE 01] LEE B., LEONG K.H., HERSZBERG I., "Effect of weaving on the tensile properties of carbon fibre tows and woven composites", *Journal of Reinforced Plastics and Composites*, vol. 20, pp. 652–670, 2001.

[LI 14] LI L., LOMOV S.V., YAN X. *et al.*, "Cluster analysis of acoustic emission signals for 2D and 3D woven glass/epoxy composites", *Composite Structures*, vol. 116, pp. 286–299, 2014.

[LI 15] LI L., LOMOV S.V., YAN X., "Correlation of acoustic emission with optically observed damage in a glass/epoxy woven laminate under tensile loading", *Composite Structures*, vol. 123, pp. 45–53, 2015.

[LI 16] LI L., SWOLFS Y., STRAUMIT I. *et al.*, "Cluster analysis of acoustic emission signals for 2D and 3D woven carbon fiber/epoxy composites", *Journal of Composite Materials*, vol. 50, pp. 1421–1435, 2016.

[LOM 02] LOMOV S.V., BELOVA E.B., BISCHOFF T. *et al.*, "Carbon composites based on multiaxial multiply stitched preforms. Part 1. Geometry of the preform", *Composites: Part A*, vol. 33, pp. 1171–1183, 2002.

[LOM 08a] LOMOV S.V., IVANOV D.S., TRUONG T.C. *et al.*, "Experimental methodology of study of damage initiation and development in textile composites in uniaxial tensile test", *Composites Science and Technology*, vol. 68, pp. 2340–2349, 2008.

[LOM 08b] LOMOV S.V., IVANOV D.S., VERPOEST I. *et al.*, "Full-field strain measurements for validation of meso-FE analysis of textile composites", *Composites: Part A*, vol. 39, pp. 1218–1231, 2008.

[LOM 09] LOMOV S.V., BOGDANOVICH A.E., IVANOV D.S. *et al.*, "A comparative study of tensile properties of non-crimp 3D orthogonal weave and multi-layer plain weave E-glass composites. Part 1: Materials, methods and principal results", *Composites: Part A*, vol. 40, pp. 1134–1143, 2009.

[LOM 15] LOMOV S.V., XU J., "Modelling high cycle fatigue of textile composites on the unit cell level", in CARVELLI V., LOMOV S.V. (eds), *Fatigue of Textile Composites*, Woodhead Publishing Limited, Cambridge, 2015.

[MOH 01] MOHAMED M.H., BOGDANOVICH A.E., DICKINSON L.C. *et al.*, "A new generation of 3D woven fabric preforms and composites", *SAMPE Journal*, vol. 37, pp. 8–17, 2001.

[MOU 07] MOURITZ A.P., "Compression properties of z-pinned composite laminates", *Composites Science and Technology*, vol. 67, pp. 3110–3120, 2007.

[MOU 08] MOURITZ A., "Tensile fatigue properties of 3D composites with through-thickness reinforcement", *Composites Science and Technology*, vol. 68, pp. 2503–2510, 2008.

[MOU 10] MOURITZ A.P., COX B.N., "A mechanistic interpretation of the comparative in-plane mechanical properties of 3D woven, stitched and pinned composites", *Composites: Part A*, vol. 41, pp. 709–728, 2010.

[MUN 02] MUNGALOV D., BOGDANOVICH A., Automated 3-D braiding machine and method, Patent No. 6,439,096, USA, 2002.

[MUN 04] MUNGALOV D., BOGDANOVICH A., "Complex shape 3-D braided composite performs: structural shapes for marine and aerospace", *SAMPE Journal*, vol. 40, pp. 7–21, 2004.

[MUN 07] MUNGALOV D., DUKE P., BOGDANOVICH A., "High performance 3-D braided fiber preforms: design and manufacturing advancements for complex composite structures", *SAMPE Journal*, vol. 43, pp. 53–60, 2007.

[QUA 14] QUARESIMIN M., CARRARO P.A., "Damage initiation and evolution in glass/epoxy tubes subjected to combined tension–torsion fatigue loading", *International Journal of Fatigue*, vol. 63, pp. 25–35, 2014.

[RUD 03] RUDOV-CLARK S., MOURITZ A.P., LEE L. *et al.*, "Fibre damage in the manufacture of advanced three-dimensional woven composites", *Composites: Part A*, vol. 34, pp. 963–970, 2003.

[SHI 96] SHIOYA M., HAYAKAWA E., TAKAKU A., "Non-hookean stress-strain response and changes in crystallite orientation of carbon fibres", *Journal of Materials Science*, vol. 31, pp. 4521–4532, 1996.

[SIM 03] SIMS G.D., "Fatigue test methods, problems and standards", in HARRIS B. (ed), *Fatigue in Composites: Science and Technology of the Fatigue Response of Fibre-Reinforced Plastics*, Woodhead Publishing Limited, Cambridge, 2003.

[SUT 09] SUTTON M.A., ORTEU J.J., SCHREIER H.W., *Image Correlation for Shape, Motion and Deformation Measurements: Basic Concepts, Theory and Applications*, Springer Science, New York, 2009.

[TAL 81] TALREJA R., "Fatigue of composite materials: damage mechanisms and fatigue-life diagrams", *Proceedings of the Royal Society of London A*, vol. 378, pp. 461–475, 1981.

[TAL 85] TALREJA R., "A conceptual framework for the interpretation of fatigue damage mechanisms in composite materials", *Journal of Composites Technology and Research*, vol. 7, pp. 25–29, 1985.

[TAL 08] TALREJA R., "Damage and fatigue in composites – a personal account", *Composites Science and Technology*, vol. 68, pp. 2585–2591, 2008.

[TAL 12] TALREJA R., SINGH C.V., *Damage and Failure of Composite Materials*, Cambridge University Press, Cambridge, 2012.

[TAL 15] TALREJA R., "A conceptual framework for studies of durability in composite materials", in CARVELLI V., LOMOV S.V. (eds), *Fatigue of Textile Composites*, Woodhead Publishing Limited, Cambridge, 2015.

[TRU 05] TRUONG CHI T., VETTORI M., LOMOV S.V. *et al.*, "Carbon composites based on multiaxial multiply stitched preforms. Part 4: Mechanical properties of composites and damage observation", *Composites Part A*, vol. 36, pp. 1207–1221, 2005.

[VAL 07] VALLONS K., ZONG M., LOMOV S.V. *et al.*, "Carbon composites based on multi-axial multi-ply stitched preforms – Part 6. Fatigue behaviour at low loads: stiffness degradation and damage development", *Composites: Part A*, vol. 38, pp. 1633–1645, 2007.

[VAL 09] VALLONS K., LOMOV S.V., VERPOEST I., "Fatigue and post-fatigue behaviour of carbon/epoxy non-crimp fabric composites", *Composites: Part A*, vol. 40, pp. 251–259, 2009.

[VAL 15] VALLONS K., "Databases for fatigue analysis in composite materials", in CARVELLI V., LOMOV S.V. (eds), *Fatigue of Textile Composites*, Woodhead Publishing Limited, Cambridge, 2015.

[VAN 02] VAN PAEPEGEM W., DEGRIECK J., "A new coupled approach of residual stiffness and strength for fatigue of fibre-reinforced composites", *International Journal of Fatigue*, vol. 24, pp. 747–762, 2002.

[VAN 05] VAN PAEPEGEM W., DEGRIECK J., "Simulating damage and permanent strain in composites under in-plane fatigue loading", *Computers & Structures*, vol. 83, pp. 1930–1942, 2005.

[VAN 15] VAN PAEPEGEM W., "Fatigue models for woven textile composite laminates", in CARVELLI V., LOMOV S.V. (eds), *Fatigue of Textile Composites*, Woodhead Publishing Limited, Cambridge, 2015.

[VAS 10] VASILOPOULOS A.P. (ed), *Fatigue Life Prediction of Composites and Composite Structures*, Woodhead Publishing Limited, Cambridge, 2010.

[YU 15] YU B., BRADLEY R.S., SOUTIS C. *et al.*, "2D and 3D imaging of fatigue failure mechanisms of 3D woven composites", *Composites: Part A*, vol. 77, pp. 37–49, 2015.

2

Fatigue Limit: A Link to Quasi-Static Damage?

2.1. Fatigue limit

The "fatigue limit" designates a load value below which fatigue damage will never happen, or will be very unlikely. Practically this definition does not really apply to composite materials, for which accelerated fatigue tests are not possible because the temperature increase for the polymer matrix places a limit for the test frequency. With the frequency range of 2–10 Hz, stipulated by the standards [AST 12], testing up to 10^6 cycles lasts 6–28 h, while testing up to 10^7 cycles lasts 60–280 h, which makes fatigue testing in order to determine the fatigue limit a very challenging endeavor. Therefore, a "fatigue limit" is most often redefined to represent a load value below which the fatigue damage at a given (large) number of cycles does not happen. The number of cycles in question is selected based on a compromise between the requirements of the application and practical limitations of the test program. The most often used value is 10^6 cycles ("operational definition") [TAL 12]. This fatigue limit will be called the "1M" fatigue limit in this chapter for brevity.

When discussing the fatigue limit in textile composites in this chapter, we restrict ourselves to the case of tension–tension fatigue with a constant (applied load) R-ratio of 0.1. We do not have enough experimental data for more complex loading cases to be able to formulate hypotheses, nor reach conclusions. We also limit the loading conditions, considering only “fiber direction” loading of a textile composite. This term will be used to designate the direction of the tension (the direction along the tensile specimen), which coincides with the direction of one of the fiber (yarn) systems of the textile reinforcement. For example, we will classify the following loading cases as “fiber direction”: tension along the warp direction (designated as 0°) of a woven laminate $[0°/90°]_s$ (but not the tension in 45° direction); tension along the direction of 60° of a braided composite with the yarns orientation [0°/+60°/–60°] (but not the tension in 90° direction); tension along the direction of 45° of a quasi-isotropic non-crimp fabric (NCF) composite $[0°/90°/–45°/+45°]_s$ (but not the tension in 22.5° direction).

Figure 2.1 illustrates the notion of the fatigue limit for the characteristic cases of glass and carbon fiber reinforced textile composites. The examples used in this figure are 3D orthogonally woven carbon and glass fiber reinforced epoxy composites studied in [CAR 10a] and [KAR 11]. Both reinforcements have the same weave structure (shown in Figure 2.1) and similar fiber volume fractions (49% and 51%, respectively). Figure 2.1 shows S–N curves for tension–tension fatigue with $R = 0.1$, loading in the warp direction. The test was load controlled. To facilitate the comparison of different materials and to highlight the mechanisms of fatigue deformation, we follow the proposal of [TAL 12] and show S–N curves using maximum strain in the first cycle as the load characterization, instead of the maximum stress value.

Three characteristic regions are marked on the S-N curves, following the theory of [TAL 12]: “static region” I, corresponding to immediate fiber breakage; “progressive region” II, corresponding to the gradual development of the matrix damage leading to the final fiber failure and “fatigue limit region” III. The 1M fatigue limit is 0.3% strain for the glass fiber reinforced and 0.8% strain for the carbon fiber reinforced 3D woven composite. The runout points on S–N diagrams (tests stopped without breakage of the specimen) seem to confirm that specimens will not fail at these loading levels for $5–10 \times 10^6$ cycles.

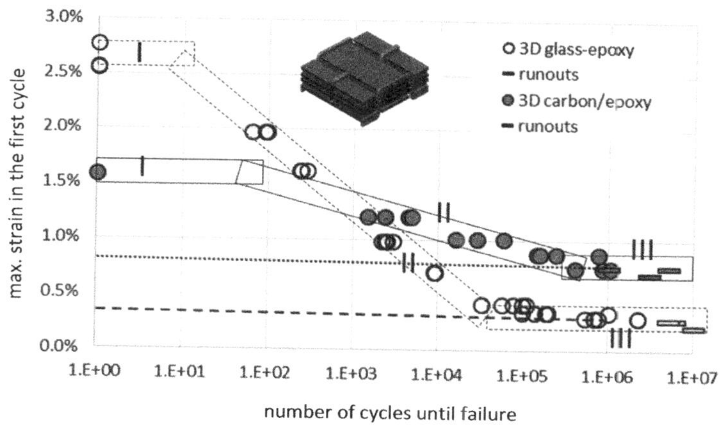

Figure 2.1. *Typical S–N curves for fiber direction loading of glass- and carbon-reinforced textile composites (inset: the weave structure for both the materials) and the "operational" fatigue limits (dashed line for glass and dotted line for carbon-reinforced composites). The data are taken from [CAR 10a] (3D woven glass-epoxy composites) and [KAR 11] (3D woven carbon/epoxy composite)*

In relation to the fatigue limit of cross-ply and multidirectional laminates, [TAL 12] makes a note which is very relevant to the question asked in the title of this chapter:

> "*Where is the fatigue limit? ... To determine the fatigue limit, if not available from the test data, a good estimate will be given by the strain at which transverse cracking initiates. The reasoning is simply that if no cracks initiate in the transverse plies, no progression of crack multiplication and subsequent damage is possible*" [TAL 12, p. 262].

From a practical point of view such estimation is extremely valuable as it allows us to find the fatigue limit without exhausting and costly test programs, or at least design such a program in an economical way, without wasting resources by narrowing the search range. The purpose of this chapter is to challenge this idea for textile composites. The fatigue life limit for short fiber reinforced composites will be discussed in Part 2 of this book.

2.2. Damage development stages and load thresholds for quasi-static tension

The damage in textile composites under quasi-static tension undergoes distinct stages, which can be characterized with the applied strain thresholds [GOR 16].

The first damage in the composite typically appears in the form of transverse (in relation to the loading direction) cracks inside the yarns or on the yarn boundaries (short designation: *t*). These cracks are developed by coalescence of the initial microdebondings on the fiber–matrix interface. Upon further load increase, the transverse cracks propagate along the yarn length. They also increase in number until a certain critical number is reached (saturation). Transverse cracks in resin-rich pockets (*tm*) are usually cracks that originated inside yarns but then propagated into these areas. When the transverse cracks are well developed, they cause the appearance of local delaminations (*l*), triggered by shear stresses resulting from the interaction of transverse cracks with longitudinal layers. Longitudinal yarns experience tensile transverse stresses caused by the Poisson's effect due to the constraint by the transverse yarns. When the transverse strength of these longitudinal yarns is exceeded by the local stress, they can start splitting (*sp*). The local delaminations progress, leading to larger inter-ply delaminations (*L*). The onset and propagation of delaminations between plies is dependent on the interlaminar fracture toughness of the composite. In the final stage, massive breakage of fibers in longitudinal yarns begins (*f*). The strain at which this happens in textile composites is typically below the ultimate strain of fibers. This reduction is caused by the fiber crimp and developed delaminations, which prohibit efficient stress transfer inside fiber bundles.

Figure 2.2(a) presents this sequence of damage events. It is illustrated (Figure 2.2(b)) by images of a transparent woven glass fiber/epoxy laminate under tension loading at different applied strains [LI 15].

The sequence of damage events suggests the presence of two thresholds of the applied load. The first, designated as ε_1, corresponds to the onset of the transverse cracking (*t*-cracks), which may not at this stage span the whole width of the specimen, being limited by the yarn crimp and/or presence of stitching in the textile reinforcement structure. The second, designated as ε_2, corresponds to, on the one hand, the onset of local delaminations (*l*-cracks) and, on the other, the formation of “strong” transverse cracks, which span the width of the specimen. Following from the detailed observations in [LI 15], in the example presented in Figure 2.2 the thresholds are as follows: ε_1 = 0.30% and ε_2 = 0.38%.

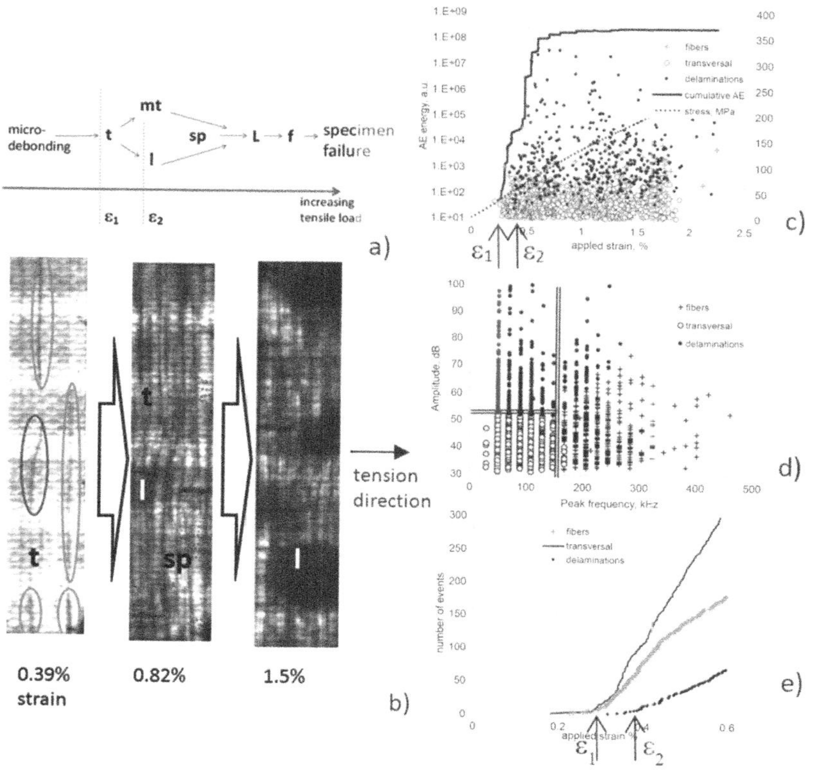

Figure 2.2. *Damage thresholds in a textile composite under tension in fiber direction: a) schematics of the damage development; b–e) example: woven glass/epoxy composite [LI 15]: b) damage seen in the backlight for several loading stages; c) stress–strain diagram with AE registration with indication of the damage thresholds; d) clustering of the AE events; e) AE event numbers in clusters as a function of the applied strain and identification of the damage thresholds*

The damage thresholds can be identified using acoustic emission (AE) registration during tensile loading. In [TRU 05, LOM 08], it was proposed to use a curve of cumulative energy (E) of the AE events for identification of the damage thresholds. As illustrated in Figure 2.2(c), the damage threshold ε_1 is related to the first increase in the $E(\varepsilon)$ curve, and the damage threshold ε_2 is related to the second sharp increase in this curve. The reader is referred also to [LOM 09] for details of the procedure of threshold identification using logarithmic and linear scales for $E(\varepsilon)$ curve. This approach was used between 2005–2015 for various textile composites; see the detailed references later in this chapter. However, the use of cumulative AE energy remains a heuristic approach for damage threshold identification.

Fibers/ matrix	Reinforcement type	Reference	ε_1, %	ε_2, %
Carbon/ epoxy	UD laminate $[0°/90°]_s$	[YOK 07]	0.42*	n/a
	UD laminate $[0°/90°]_s$	[TAL 12]	0.40*	n/a
	UD $[0°/90°/+45°/–45°]_s$	[TAL 12]	0.59*	n/a
	NCF $[0°/90°]_s$	[TRU 05]	0.30	0.62
	NCF $[0°/90°/+45°/–45°]_s$	[MIK 08]	0.40	0.60
	Woven twill 2/2 laminate	[DE 11]	0.26	0.34
	braided $[0°/+45°/–45°]_4$	[IVA 09b]	0.29	0.55
	3D orthogonally woven	[BOG 13]	0.64	0.72
Glass/epoxy	Woven plain weave laminate	[LI 15]	0.30	0.42
	Woven plain weave laminate	[LOM 09]	0.26	0.43
	3D orthogonally woven	[LOM 09]	0.43	0.54

*Based on the crack density measurements.

Table 2.1. *Examples of the damage threshold values (loading in 0°direction)*

Recent research on identification of the damage modes in fiber-reinforced composites using cluster analysis added frequency and other descriptors of the AE events to the damage identification descriptors (for example [GUT 11, SAU 12, BAK 15]). In [LI 14, LI 15, LI 16], this approach was applied to textile composites, which allowed more physically based identification of the damage thresholds. In brief, the cluster analysis of the AE events typically identifies three clusters of events in the multiparametric event space. The most important event parameters are the AE signal amplitude (expressed in dB and strongly correlated with the logarithm of the AE event energy) and either of the frequency-related parameters (peak frequency or frequency centroid). The three clusters, shown in Figures 2.2(c) and (d), contain events originated from transverse matrix cracks, local delaminations and fiber breakage. The load values, corresponding to the onset of the AE events in the first and the second cluster, correspond to the damage threshold values ε_1 and ε_2, respectively. The reader is referred to [LI 14, LI 15, LI 16] for details of the cluster analysis procedure and experimental validation of this approach.

Table 2.1 gives examples of the damage threshold values for different textile composites, measured using AE registration. For comparison, the values of the transverse cracking initiation for unidirectional laminates, based on the counting of direct cracks, are given. The damage thresholds for typical textile composites (apart from 3D orthogonally woven composites with exceptionally straight yarns and well-aligned fibers) lie in the range ε_1 = 0.25–0.4% and ε_2 = 0.35–0.6%. The ε_1 threshold for textile composites is significantly lower than for UD laminates.

The cluster-based identification of damage thresholds does not contradict the previously used approach based on cumulative energy curves, as the main distinctive parameter between the transverse cracking and local delaminations is the event amplitude or its acoustic energy. Therefore, the previous data collected using cumulative energy curves preserves its validity. In the new research, cluster analysis identification of damage thresholds is recommended as being more physically sound.

2.3. Damage development in quasi-static tension and in the progression of fatigue loading

The development of damage during tension–tension fatigue loading of textile composites with brittle matrices is analogous to the sequence of the damage event outlined in the previous section for quasi-static loading. This is illustrated in Figure 2.3. It shows a sequence of damage events in 3D woven glass/epoxy composites [IVA 09a, CAR 10a] (an S–N curve for the same material is shown in Figure 2.1) for three loading cases, all of them with the loading aligned with the warp direction: quasi-static tension, tension–tension fatigue with R = 0.1 at the maximum load corresponding to 1.0% (fatigue failure at 2,000 cycles) and 0.25% strain in the first cycle (no fatigue failure up to 5,000,000 cycles).

The sequence of the crack development under fatigue loading is very much the same as under quasi-static tension: initiation, growth and multiplication of the transverse cracks. In the case of the high fatigue load (1% maximum strain), the development of the transverse cracks system leads to local delaminations, associated with the positions of the "crowns" of the binding Z-yarns in the 3D weave. For lower load (0.25% maximum strain), these delaminations do not appear.

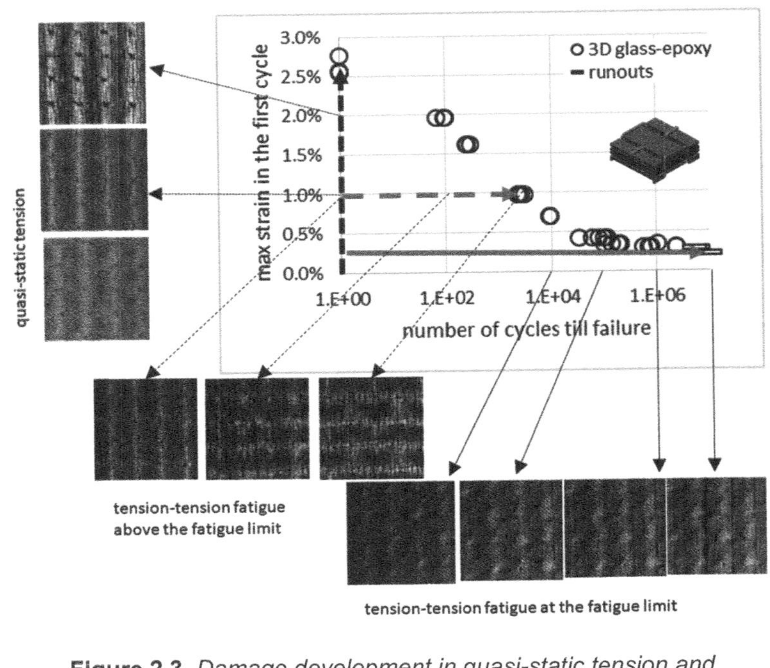

Figure 2.3. *Damage development in quasi-static tension and tension–tension fatigue. 3D woven glass-epoxy composites, loading in warp direction (horizontal direction of the images) [IVA 09a, CAR 10a]*

The driving force of the damage development under quasi-static loading is the increasing load and corresponding influx of the deformation energy, which can be used for, and released during, the new cracks' formation and propagation of the existing cracks. The "decision", if a new crack will appear and if the existing crack will propagate, is defined by the level of local stress. Under fatigue loading, the deformation energy is supplied again and again under loading cycles. The appearance and propagation of the cracks is controlled by the local fatigue strength of the matrix under the local stress state defined by the maximum of the loading cycle.

Figure 2.4 shows an evolution of the crack density during tension–tension fatigue loading ($R = 0.1$) of carbon/epoxy composites reinforced with biaxial NCF [+45°/–45°] [VAL 09]. The tension is applied in the direction of the fibers (in +45° and –45° directions, see the two curves on the graph) with the maximum strain in the first cycle of 0.8%. This value is much higher than the damage initiation strain in quasi-static test (0.36–0.40%), and yet it does not lead to fatigue failure, therefore corresponding to the fatigue limit.

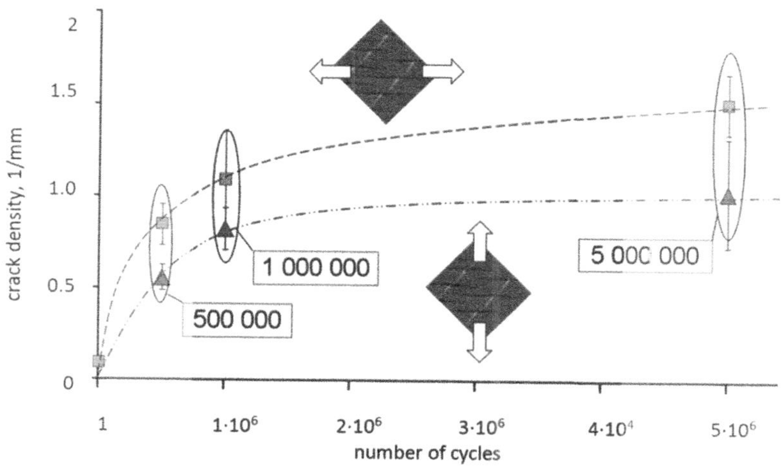

Figure 2.4. *Crack density during tension–tension fatigue loading of carbon/ epoxy NCF [+45°/–45°] composites in the fiber direction [VAL 09]*

The saturation behavior of the crack density curves is quite similar to the saturation of the transverse crack system during quasi-static loading. It takes more than one million cycles to reach saturation; this value, as well as the range of the crack densities, corresponds to the observations reported in [TAL 12] for tension–tension cross-ply laminates with unidirectional layers. The range 1.0–1.5 1/mm also supports the crack density values attained in the quasi-static tension of cross-ply laminates.

2.4. Experimental data on the fatigue limit and the quasi-static damage thresholds for textile composites

The approach to studies of the mechanical behavior and the microstructure of textile composites, nicknamed a "Chinese Puzzle" in [LOM 14], constitutes an integrated research methodology of textile composites, when the properties of the same material are studied through all the stages of its production and performance (Figure 2.5). This includes the damage and strength of the material under quasi-static and fatigue loading.

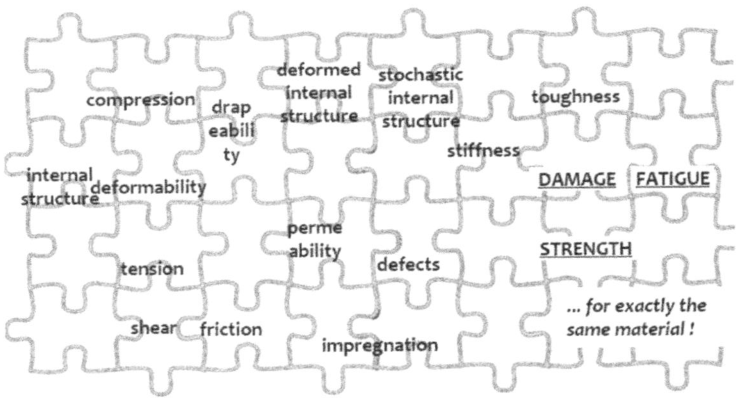

Figure 2.5. *A "Chinese puzzle" for integrated research of textile composites [LOM 14]*

Over the last decade, the authors of this book have applied this methodology to a broad range of textile composites with 2D and 3D woven, braided, NCF and structurally stitched architectures of glass and carbon fiber reinforcements. This work has created a dataset that includes damage thresholds under quasi-static tension and S–N curves and fatigue limits under tension–tension fatigue. Tables 2.2 and 2.3 present this dataset for glass (Table 2.2) and carbon (Table 2.3) fiber reinforced thermoset materials with different reinforcement architectures. For each material, a short description of the reinforcement, fibers, fiber volume fraction and matrix is provided, together with the reference to the original publication where the reader is referred to for full details of the materials, experimental procedure, data reduction and the statistical treatment.

All the composites presented in Tables 2.2 and 2.3 have a fiber volume fraction in the range of 55–65%. As mentioned in the introduction of this chapter, loading is the fiber direction tension. The exception to this is a 3D braided composite with a braiding angle of 10°, loaded in 0° direction; the damage mechanisms in this case are close to the ones during the fiber direction tension. For non-symmetric reinforcement geometry, the data for loading in two orthogonal directions are given.

Loading, both quasi-static and fatigue, is expressed in terms of the applied strain. For quasi-static loading, the strain was measured using mechanical or optical extensometers. For tension–tension fatigue tests, performed on stress control hydraulic machines, the value of the strain, characterizing the loading, corresponds to the maximum strain in the first fatigue cycle. All fatigue tests were done with $R = 0.1$, frequency from 2 to 10 Hz, and with an appropriate temperature control, which ensured that the temperature rise during the fatigue loading was within 10°C.

The main focus of this chapter is the values of the damage thresholds and 1M fatigue limit. The values for damage thresholds are given as mean values and standard deviations in at least five tests in each case. The values of 1M fatigue limit are defined using a three-linear representation of an S–N curve in semi-logarithmic co-ordinates (see Figure 2.1). The scatter of the 1M fatigue limit strain value, determined in this manner, is about 0.05% strain; only mean values are given in Tables 2.2 and 2.3. Apart from the values for the damage thresholds and the 1M fatigue limit, an S–N curve is given for each material. The S–N graphs show measured experimental points and give an idea of the experimental scatter of the 1M fatigue limit in each case.

The mean values of the damage thresholds ε_1 and ε_2 are shown as dotted (ε_1) and dashed (ε_2) lines on the S–N curves. This allows for a visual comparison of the damage thresholds and the fatigue limit.

2.4.1. *Fatigue limit for glass fiber reinforced composites*

The data for glass fiber reinforced composites is shown in Table 2.2. It is limited to a plain weave laminate and 3D orthogonally woven reinforcement. The values of the 1M fatigue limit in both the cases lie in the range of 0.30–35% strain. These values are much lower than the 1M fatigue limit of unidirectional glass bundles; for the latter, different sources give values in the range of 0.8–1.5% [TAL 12, ABD 15]. This is not surprising, as fiber crimp and interaction between the yarns in a textile structure create a complex stress state in the yarns and fibers.

The 1M fatigue limit ε_{1M} is well below the ε_2 quasi-static damage threshold, with the difference $\varepsilon_2 - \varepsilon_{1M}$ in the range of 0.1–0.2% strain. This difference may seem small, but it corresponds to 30–60% of the ε_{1M} value. This means that the immediate development of strong damage modes as long transverse cracks and local delamination is not a necessary prerequisite for fatigue failure. The latter can develop even in the presence of smaller transverse cracks, which are created after the ε_1 threshold is reached in the first cycle loading.

The 1M fatigue limit ε_{1M} is close to the ε_1 threshold. This closeness seems to confirm the logic of [TAL 12] cited in section 2.1: "... *if no cracks initiate in the transverse plies, no progression of crack multiplication and subsequent damage is possible*". However, this simple statement is not true for textile composites. Figure 2.3 shows the development of fatigue damage in the same 3D orthogonally woven glass fiber composite shown in Table 2.2 at the maximum strain level 0.25%. We see that, in spite of low maximum strain in the first cycle, which, being below the damage initiation threshold of 0.43 ± 0.04%, cannot lead to the appearance of transverse cracks in weft yarns (consider warp loading), the transverse cracks appear with the progression of the fatigue cycles. This may be a result of fatigue-induced cracking of the matrix, which can happen on lower stress/strain levels than cracks during quasi-static loading.

Material description and load direction, reference	Reinforcement structure	Fibers (volume fraction) and matrix	Quasi-static damage thresholds strain, %		S–N diagram, tension–tension fatigue, $R = 0.1$ and quasi-static damage thresholds (dotted (ε_1) and dashed (ε_2) lines)	1M fatigue limit, max. strain first cycle, %
			ε_1	ε_2		
2D plain weave laminate load in warp direction [IVA 09a, LOM 09, CAR 10a]		Hybon 2022 E-glass (52%) Derakane 8084 epoxy-vinyl ester	0.26 ± 0.04	0.43 ± 0.06	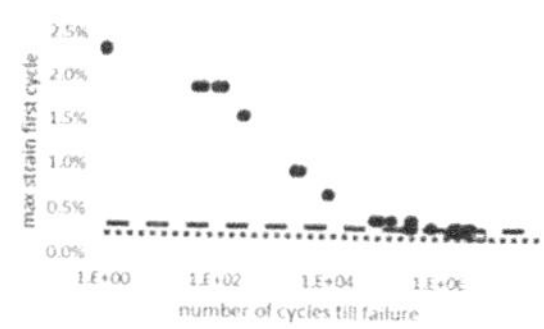	0.34
3D orthogonal weave load in warp/weft direction [IVA 09a, LOM 09, CAR 10a]		Hybon 2022 E-glass (52%) Derakane 8084 epoxy-vinyl ester	0.43 ± 0.04 0.37 ± 0.06	0.54 ± 0.04 0.59 ± 0.04	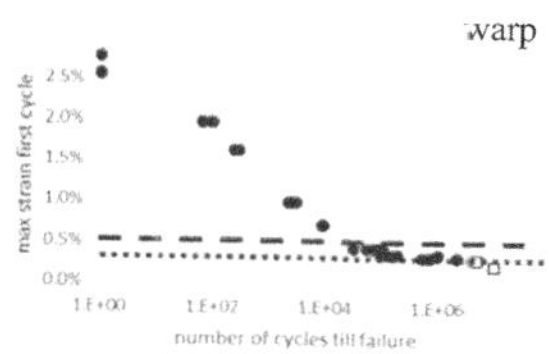	0.31 0.35

Runouts are shown on S–N curves with open symbols.

Table 2.2. *Tension–tension fatigue data in comparison with the quasi-static damage thresholds. Glass-reinforced composites*

We can conclude that for glass reinforced textile composites, the 1M fatigue life strain is close to the ε_1 quasi-static damage threshold. This seems valid for the two composites investigated for which the 1M fatigue life limits and damage thresholds differ by 1.5–2 times; this insensitivity to the material type gives certain credibility to the conclusion.

2.4.2. Fatigue limit for carbon fiber reinforced composites

The data for carbon fiber reinforced composites are shown in Table 2.3. It includes a large variety of reinforcement structures: 2D and 3D woven, NCF and braided. The 1M fatigue life limit strain in all the cases is above 0.5% and can be as high as 1.1%.

On the one hand, in all the cases, the 1M fatigue life limit strain is higher than the ε_1 and ε_2 damage threshold, in most cases much higher. It seems that the fatigue life is controlled by the fatigue of the carbon fibers themselves, rather than the development of matrix damage. Comparison between the cases of the stitched and unstitched NCF gives an illustration for this hypothesis. Both damage thresholds for the structurally stitched material are much lower than the thresholds for the unstitched, especially ε_2 (0.22–0.26% for two loading directions in the stitched case and 0.45–0.57% unstitched). The difference of the damage threshold and the whole process of the damage development in quasi-static loading is attributed in [CAR 10] to the fiber misalignment and resin-rich zones created by structural stitching and the related stress concentrations. However, the 1M fatigue life limits for these two materials are the same (0.60–0.62%).

Material description and load direction, reference	Reinforcement structure	Fibers (volume fraction) and matrix	Quasi-static damage thresholds strain, %		S–N diagram, tension–tension fatigue, $R = 0.1$ and quasi-static damage thresholds (dotted (ε_1) and dashed (ε_2) lines)	1M fatigue limit, max. strain first cycle, %
			ε_1	ε_2		
2D twill 2/2 laminate load in warp direction [XU 14, XU 16]	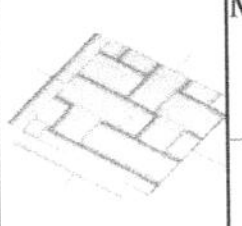	T700S 3K (55%), epoxy M10.1	0.30 ± 0.05	0.35 ± 0.06	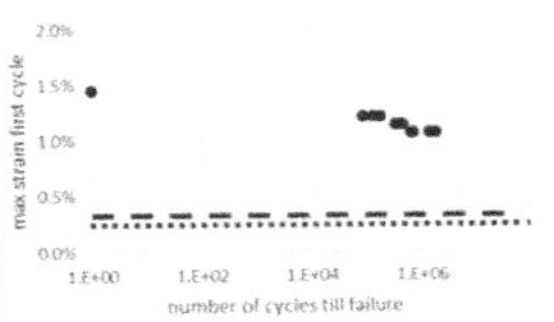	1.06
		T700S 12K (53%), epoxy M10.1	0.42 ± 0.04	0.48 ± 0.05	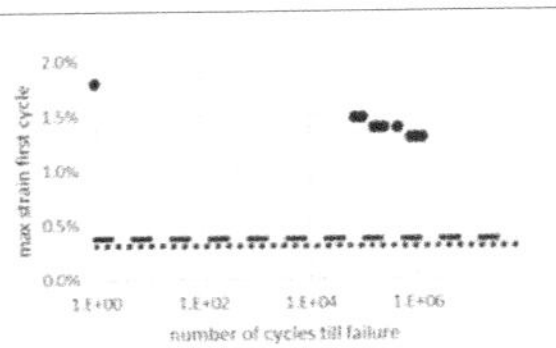	1.30

Material description and load direction, reference	Reinforcement structure	Fibers (volume fraction) and matrix	Quasi-static damage thresholds strain, %		S–N diagram, tension–tension fatigue, $R = 0.1$ and quasi-static damage thresholds (dotted (ε_1) and dashed (ε_2) lines)	1M fatigue limit, max. strain first cycle, %
			ε_1	ε_2		
3D orthogonal weave load in warp/weft direction [KAR 11, BOG 13]		Toho Tenax HTS40 (51%) epoxy West System 105, 209 hardener	0.64 ± 0.07 0.61 ± 0.14	0.72 ± 0.13 0.71 ± 0.15	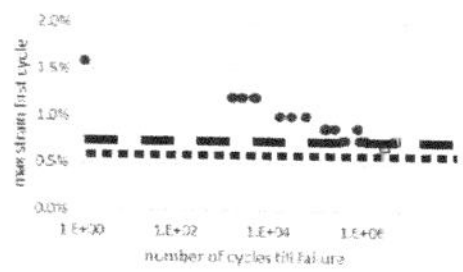	0.80 0.76
[+45/–45]$_{2S}$ NCF load in +45°/–45° direction [VAL 09]		24K carbon HTS (58%), epoxy	0.40 ± 0.05 0.36 ± 0.03	0.62 ± 0.05 0.60 ± 0.05	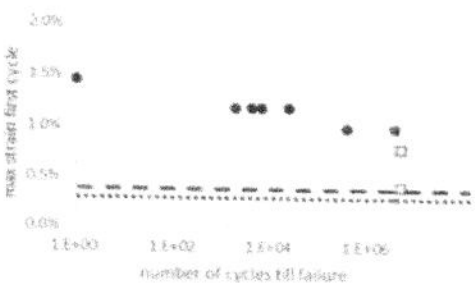	1.00 1.10
[45/–45/0/90/45/–45]$_S$ NCF load in 0°/90° direction [CAR 09, CAR 10b]	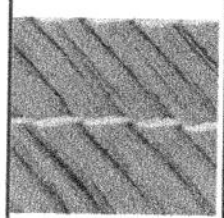	24K Tenax HTS 5631 (63%), epoxy HexFlow RTM-6	0.16 ± 0.10 0.13 ± 0.10	0.57 ± 0.05 0.45 ± 0.09	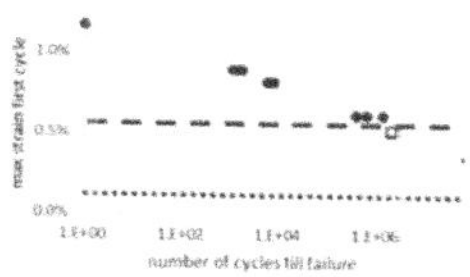	0.61 0.61
[45/–45/0/90/45/–45]$_S$ structurally stitched NCF load in 0°/90° direction [CAR 09, CAR 10b]		24K Tenax HTS 5631 (63%), stitching Tenax HTA 5241, epoxy HexFlow RTM-6	0.12 ± 0.04 0.08 ± 0.04	0.26 ± 0.06 0.22 ± 0.06	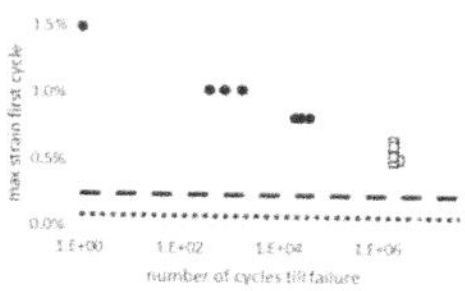	0.60 0.61
3D braid, braiding angle 10°, load in 0° direction, [CAR 2013]	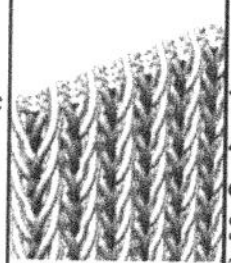	12K Toho Tenax HTS 40 (56%), epoxy West System 105, 209 hardener	0.12 ± 0.03	0.37 ± 0.08	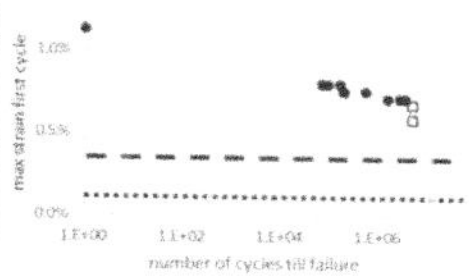	0.72

Runouts are shown on S–N curves with open symbols.

Table 2.3. *Tension-tension fatigue data in comparison with the quasi-static damage thresholds. Carbon reinforced composites*

On the other hand, the 1M fatigue life limit never gets below the ε_2 threshold, but the 3D, orthogonally woven reinforcement and structurally stitched NCF are very close to it. The values of the 1M fatigue life limit in the range of 0.5–1.1% are significantly lower than the values for the single fiber or unidirectional fiber bundle test, which are in the range of 80–98% of the ultimate tensile strain of the fibers, hence about 1.8–2.0% [ABD 15]. Crimp of the fibers in the textile structure definitely plays a role in this reduction of the fatigue life limit. When the ε_2 threshold is reached, local delaminations appear in the textile structure that may create possibilities for fibers and fiber bundles to bend and to be subjected to a combination of tension and shear, which decreases the fatigue resistance.

In conclusion, the quasi-static damage threshold ε_2 can be considered as a lower bound estimation of the 1M fatigue life limit for carbon fiber reinforced composites in tension–tension (R = 0.1) fatigue loading in fiber direction. This estimation may be quite conservative in some cases (with the fatigue life limit 1.5–2 times higher than ε_2), but there are cases when the two values are very close.

2.5. Summary and conclusion on the fatigue life limit

We have presented a dataset, describing quasi-static and tension–tension fatigue behavior of various glass and carbon fiber reinforced thermoset textile composites. The dataset spans woven (2D and 3D), multiaxial warp knitted and braided reinforcements. For all the materials, the damage thresholds in quasi-static tension, determined by AE analysis and confirmed by direct observation in stopped tests, were compared with tension–tension (R = 0.1) fatigue life limits, leading to the maximum strain in the first loading cycle giving the fatigue life of 1 million cycles.

We have confirmed the link between quasi-static damage thresholds and the fatigue life limit. Figure 2.6 summarizes the comparison between the fatigue life limit and the quasi-static damage thresholds under quasi-static loading (first loading cycle in fatigue): ε_1, which represents the initiation of matrix cracking in the transverse yarn system of the textile structure, and ε_2, which represents local delaminations on the scale of the reinforcement unit cell and the development of "strong" transverse cracks, which span the width of the specimen. It shows the ratio $\varepsilon_{1M\_FL}/\varepsilon_1$ (for glass fiber reinforced composites) and $\varepsilon_{1M\_FL}/\varepsilon_2$ (for carbon fiber reinforcements) for all the materials in the presented dataset.

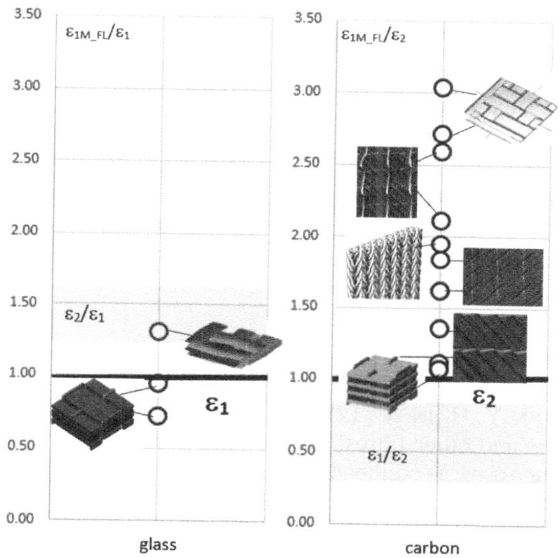

Figure 2.6. *Relation of the 1M fatigue life limit strain and damage thresholds ε_1 and ε_2 for different textile composites*

The behavior of textile composites proved to be more complex than the simple hypothesis proposed in [TAL 12] suggests, "a good estimate [of the fatigue limit] will be given by the strain at which transverse cracking initiates".

This statement was seen to be true for glass-reinforced composites. For them, the 1M fatigue life limit is close to the quasi-static damage threshold ε_1. The notable fact is that at loads below the fatigue life limit, when the sample does not fail after several million fatigue cycles, transverse matrix cracking still appears after a certain number of cycles because of matrix fatigue.

For carbon reinforced materials, the ε_1 damage threshold is a serious underestimation of the fatigue life limit. The ε_2 threshold serves better as a conservative estimation of the 1M fatigue life limit. The underestimation can be as high as three times, depending on the material (see points corresponding to the carbon twill woven composites in Figure 2.6), but for some materials the strain of the 1M fatigue life limit is very close to the ε_2 threshold.

The observed relations between the fatigue life limit and the quasi-static damage threshold provide a rough estimation of the design strains to use under fatigue strength requirements and allow planning of the fatigue testing programs to minimize the amount of costly and time-consuming experiments with low loads and high numbers of cycles.

2.6. Bibliography

[ABD 15] ABDIN Y., JAIN A., LOMOV S.V. *et al.*, "5 – Fatigue analysis of carbon, glass and other fibres", in CARVELLI V., LOMOV S.V., *Fatigue of Textile Composites*, Woodhead Publishing Limited, Cambridge, 2015.

[AST 12] ASTM, Standard test method for tension-tension fatigue of polymer matrix composite materials D3/79/D3479M Standard, 2012.

[BAK 15] BAKER C., MORSCHER G.N., PUJAR V.V. *et al.*, "Transverse cracking in carbon fiber reinforced polymer composites: modal acoustic emission and peak frequency analysis", *Composites Science and Technology*, vol. 116, pp. 26–32, 2015.

[BOG 13] BOGDANOVICH A.E., KARAHAN M., LOMOV S.V. *et al.*, "Quasi-static tensile behavior and progressive damage in carbon/epoxy composite reinforced with 3D non-crimp orthogonal woven fabric", *Mechanics of Materials*, vol. 62, pp. 14–31, 2013.

[CAR 09] CARVELLI V., KOISSIN V., KUSTERMANS J. *et al.*, "Progressive damage in stitched composites: Static tensile tests and tension-tension fatigue", *17th International Conference on Composite Materials (ICCM-17)*, IOM Communications Ltd., Edinburgh, 2009.

[CAR 10a] CARVELLI V., GRAMELLINI G., LOMOV S.V. *et al.*, "Fatigue behaviour of non-crimp 3D orthogonal weave and multi-layer plain weave E-glass reinforced composites", *Composites Science and Technology*, vol. 70, pp. 2068–2076, 2010.

[CAR 10b] CARVELLI V., TOMASELLI V.N., LOMOV S.V. *et al.*, "Fatigue and post-fatigue tensile behaviour of non-crimp stitched and unstitched carbon/epoxy composites", *Composites Science and Technology*, vol. 70, pp. 2216–2224, 2010.

[CAR 13] CARVELLI V., PAZMINO J., LOMOV S.V. *et al.*, "Quasi-static and fatigue tensile behavior of a 3d rotary braided carbon/epoxy composite", *Journal of Composite Materials*, vol. 47, pp. 3188–3202, 2013.

[DE 11] DE GREEF N., GORBATIKH L., GODARA A. *et al.*, "The effect of carbon nanotubes on the damage development in carbon fiber/epoxy composites", *Carbon*, vol. 49, pp. 4650–4664, 2011.

[GOR 16] GORBATIKH L., LOMOV S.V., "Damage accumulation in textile composites", in TALREJA R., VARNA J., *Modeling Damage, Fatigue and Failure of Composite Materials*, Elsevier, Cambridge, 2016.

[GUT 11] GUTKIN R., GREEN C.J., VANGRATTANACHAI A. *et al.*, "On acoustic emission for failure investigation in CFRP: pattern recognition and peak frequency analyses", *Mechanical Systems and Signal Processing*, vol. 25, pp. 1393–1407, 2011.

[IVA 09a] IVANOV D.S., LOMOV S.V., BOGDANOVICH A.E. *et al.*, "A comparative study of tensile properties of non-crimp 3D orthogonal weave and multi-layer plain weave E-glass composites. Part 2: Comprehensive experimental results", *Composites Part A*, vol. 40, pp. 1144–1157, 2009.

[IVA 09b] IVANOV D.S., LOMOV S.V., BAUDRY F. *et al.*, "Failure analysis of triaxial braided composite", *Composites Science and Technology*, vol. 69, pp. 1372–1380, 2009.

[KAR 11] KARAHAN M., LOMOV S.V., BOGDANOVICH A.E. *et al.*, "Fatigue tensile behavior of carbon/epoxy composite reinforced with non-crimp 3D orthogonal woven fabric", *Composites Science and Technology*, vol. 71, pp. 1961–1972, 2011.

[LI 14] LI L., LOMOV S.V., YAN X., *et al.*, "Cluster analysis of acoustic emission signals for 2D and 3D woven glass/epoxy composites", *Composite Structures*, vol. 116, pp. 286–299, 2014.

[LI 15] LI L., LOMOV S.V., YAN X., "Correlation of acoustic emission with optically observed damage in a glass/epoxy woven laminate under tensile loading", *Composite Structures*, vol. 123, pp. 45–53, 2015.

[LI 16] LI L., SWOLFS Y., STRAUMIT I. *et al.*, "Cluster analysis of acoustic emission signals for 2D and 3D woven carbon fiber/epoxy composites", *Journal of Composite Materials*, vol. 50, pp. 1421–1435, 2016.

[LOM 08] LOMOV S.V., IVANOV D.S., TRUONG C.T. *et al.*, "Experimental methodology of study of damage initiation and development in textile composites in uniaxial tensile test", *Composites Science and Technology*, vol. 68, pp. 2340–2349, 2008.

[LOM 09] LOMOV S.V., BOGDANOVICH A.E., IVANOV D.S. *et al.*, "A comparative study of tensile properties of non-crimp 3D orthogonal weave and multi-layer plain weave E-glass composites. Part 1: materials, methods and principal results", *Composites Part A*, vol. 40, pp. 1134–1143, 2009.

[LOM 14] LOMOV S.V., VERPOEST I., "Chinese puzzles of textile composites properties", *Proceedings of the 16th European Conference on Composite Materials (ECCM-16)*, Seville, Spain, 2014.

[MIK 08] MIKHALUK D.S., TRUONG T.C., BOROVKOV A.I. *et al.*, "Experimental observations and finite element modelling of damage and fracture in carbon/epoxy non-crimp fabric composites", *Engineering Fracture Mechanics*, vol. 75, pp. 2751–2766, 2008.

[SAU 12] SAUSE M.G.R., GRIBOV A., UNWIN A.R. *et al.*, "Pattern recognition approach to identify natural clusters of acoustic emission signals", *Pattern Recognition Letters*, vol. 33, pp. 17–23, 2012.

[TAL 12] TALREJA R., SINGH C.V., *Damage and Failure of Composite Materials*, Cambridge University Press, 2012.

[TRU 05] TRUONG C.T., VETTORI M., LOMOV S.V. *et al.*, "Carbon composites based on multiaxial multiply stitched preforms. Part 4: Mechanical properties of composites and damage observation", *Composites Part A*, vol. 36, pp. 1207–1221, 2005.

[VAL 09] VALLONS K., LOMOV S.V., VERPOEST I., "Fatigue and post-fatigue behaviour of carbon/epoxy non-crimp fabric composites", *Composites Part A*, vol. 40, pp. 251–259, 2009.

[XU 14] XU J., LOMOV S.V., VERPOEST I. *et al.*, "A progressive damage model of textile composites on meso-scale using finite element method: static damage analysis", *Journal of Composite Materials*, vol. 48, pp. 3091–3109, 2014.

[XU 16] XU J., LOMOV S.V., VERPOEST I. *et al.*, "A comparative study of twill weave reinforced composites under tension-tension fatigue loading: experiments and meso-modelling", *Composite Structures*, vol. 135, pp. 306–315, 2016.

[YOK 07] YOKOZEKI T., IWAHORI Y., ISHIWATA S., "Matrix cracking behaviors in carbon fiber/epoxy laminates filled with cup-stacked carbon nanotubes (CSCNTs)", *Composites Part A*, vol. 38, pp. 917–924, 2007.

PART 2

Fatigue of Short Fiber Reinforced Composites

3

Experimental Observations of Fatigue of Short Fiber Reinforced Composites

This chapter begins with a brief introduction on short fiber reinforced composites (SFRCs). For the discussion in this chapter and book, we focus entirely on injection molded short fiber composites. The fatigue behavior of such materials is also described in detail. The response of SFRC to fatigue loading manifests itself in two easily observable ways. First is the Stress-Number (SN) of Cycles curve; it represents the number of cycles to failure if a certain load is applied. Second is the loss of stiffness; SFRC materials suffer from continuous loss of stiffness during cyclic loading.

A detailed description of the SN curve behavior of SFRCs is provided in section 3.2. The dependence of the SN curve on various materials, environmental and loading factors will be elucidated upon. Next, the loss of stiffness during cyclic loading is described in section 3.3. Emphasis is given to statistical treatment of fatigue data. Statistical methods to access variance in fatigue data have been tried for metals and welds [SCH 03, SUT 00]. Proper use of statistics can provide insights into composite fatigue as well. For example, using simple statistics, Marshall *et al.* [MAR 85] were able to prove that the scatter in fatigue data of glass fiber unidirectional (UD) composites was due to change in the failure mode.

These sections will be used to highlight the key challenges and future outlook for design and simulation of fatigue behavior of SFRC explored in section 3.4.

3.1. Injection molded SFRC

Injection molding is a manufacturing process for producing parts by injecting material into a mold. Material for the part is fed into a heated barrel, mixed and forced into a mold cavity, where it cools down and consolidates to the configuration of the cavity [TOD 94]. A schematic representation of the injection molding process is shown in Figure 3.1.

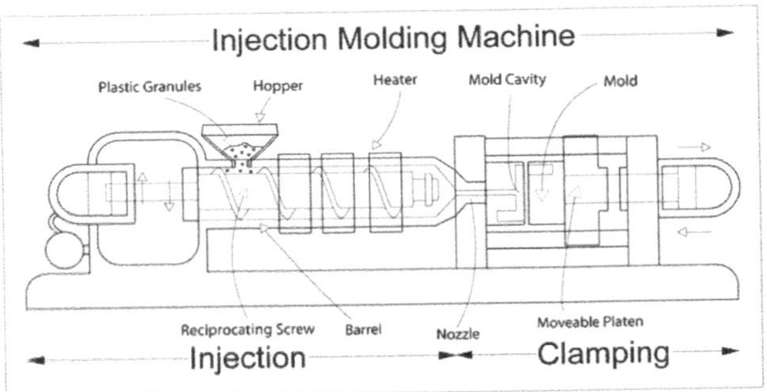

Figure 3.1. *Schematic representation of the injection molding process [UNI 16]*

Injection molded SFRCs are easy to manufacture and have reasonable specific strengths compared to conventionally used metals. Typically, SFRCs are used for semistructural components in automobiles. Some examples of SFRC components in automobiles are engine mounts, spare wheel recesses and engine cover cams.

Despite several possible advantages of utilizing SFRC, their use in the automobile industry today is rather limited. One of the main reasons for this is that the fatigue behavior of such materials is hard to model. The fatigue behavior of SFRC depends on a large number of factors and the direct correlation between the factors and fatigue properties is usually difficult to make. As a direct consequence, fast and reliable simulation methods for functional performances such as durability are not easily available. The development time of vehicles, which use conventional engineering materials, has decreased drastically. Design methodologies of SFRC materials have not yet reached similar levels of maturity. In the following sections, the different aspects of the fatigue behavior of SFRCs will be described.

Precompounded glass fiber reinforced polymer pellets are typically used for the injection molding process to manufacture SFRC components. The fiber orientation distribution (FOD) depends on how the matrix flows in the mold. Apart from the properties of the matrix, the flow of the liquid matrix depends on the injection parameters and the specific shape of the mold [CHA 00]. The FOD also varies from point to point in a component (Figure 3.2).

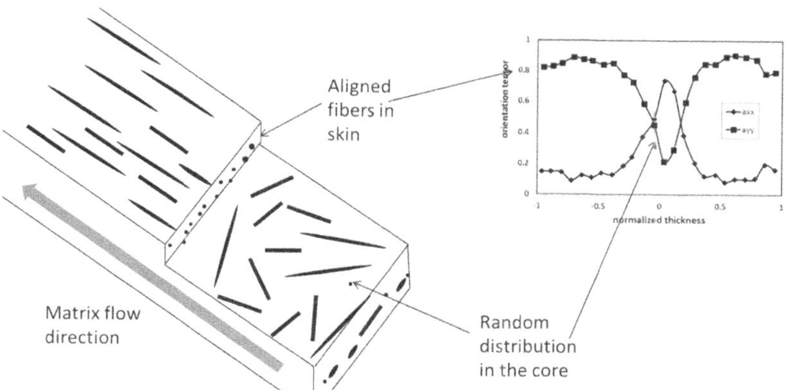

Figure 3.2. *Variation of fiber orientations in glass fiber reinforced injection molded component. On the left is a simplistic representation of the cross section showing the different fiber orientations through the thickness. On the right is the numerical representation of the FOD variation with the help of orientation tensor*

Note, the "*skin core*" distribution of the FOD in Figure 3.2. Skin core is a commonly used term to describe the variation of the FOD through the thickness of the SFRC component. This unique variation of the FOD through the thickness is caused due to the effect of shear and extensional flow. Depending on the relative position with respect to the mold wall, the effect of shear or extensional flow becomes predominant. Near the wall of the mold the FOD is random with respect to the flow direction. This region is typically very thin. The region adjacent to it is the shell region. The shell region is characterized by an FOD almost parallel to the direction of flow of the matrix (also known as matrix flow direction [MFD]). Finally, there is the core region where the extensional flow predominates and the fibers align themselves almost perpendicular to the direction of flow of the matrix [BER 08, HOR 97a, KAR 89]. Due to this skin core effect, a direct co-relation between the FOD and measured fatigue properties is not always easy to make.

The FOD of SFRC materials is usually characterized by a so-called second-order orientation tensor [ADV 87], which can be derived experimentally using micro-CT [ABD 14], scanning electron microscopy [EIK 13, GUI 93] or radiography-based characterization; alternately there are also commercial manufacturing simulation software [ZHO 13]. The algorithms behind manufacturing simulation are known to be accurate when the fiber content is low; its accuracy reduces as the fiber content increases [KIM 01].

The variation of fiber length in such materials is not usually significant. It is common to assume a single average fiber length to the entire SFRC component. The typical aspect ratio of the fibers (length/diameter) is between 15–60. A similar simplification is also made for the fiber volume fraction; it is a reasonable assumption to believe that the fiber dispersion through the volume of the component is uniform. Typically, fiber content in the range of 10–50% by weight is employed for commercial applications [PLA 14]. A wide range of fibers including glass, carbon, boron, steel and some natural fiber composites have been employed for making injection molded fiber-reinforced composites. However, the use of glass fiber is the most common with over 95% of the fiber-reinforced injection molded composites made using glass fiber as reinforcement [BIR 12]. This is probably due to its relatively low cost, reasonable strength and other attractive properties like chemical resistance and insulating capabilities. In this chapter, the fatigue behavior of glass fiber reinforced SFRC is given the most attention.

3.2. SN curve behavior of SFRC

In this section, the SN curve behavior of SFRC materials is described. A description of the experimental results describing several factors on which the fatigue behavior of SFRC depends is presented below. For each of the factors, the reported dependence of the slope and strength is recalled. Beforehand, we present a brief overview of the prevalent protocols for fatigue testing of SFRC.

Coupons for fatigue testing are commonly prepared by injection molding plates and machining coupons in different directions with respect to the MFD (see Figure 3.3).

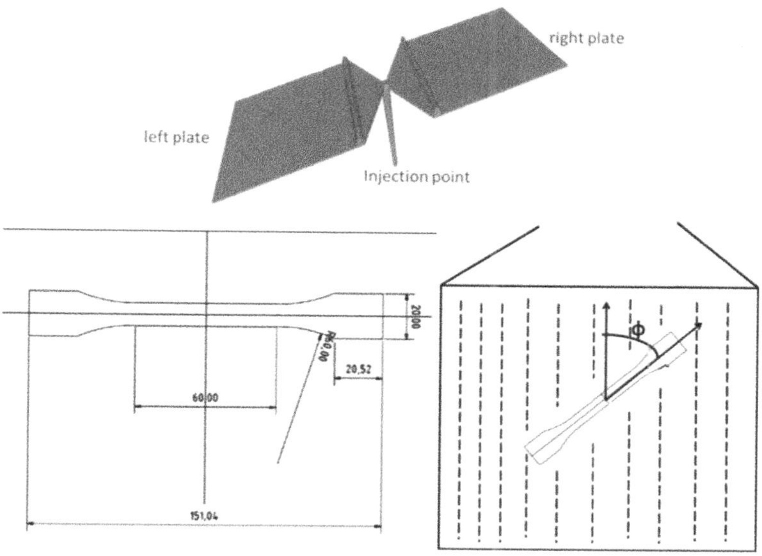

Figure 3.3. *Specimens preparation for fatigue testing of SFRC: a) plates used for injection molding; b) the geometry of the dog bone specimen with dimensions in millimeter; c) orientation designation of coupon [JAI 16c]*

This method has been used by a large number of researchers [ARI 14a, BER 07a, DE 12, HAS 12, HOR 97b, JAI 16c, KLI 11a]. Wyzgoski and Novak [WYZ 05] used ribbed plates to generate coupons for both tension and flexural fatigue testing. There is no general consistency on the gauge length or thickness of the coupon to be tested. Arif *et al.* [ARI 14a] performed fatigue tests on coupons with a gauge length of 20 mm, while Jain *et al.* [JAI 16c] had a gauge length of 60 mm. The rest of the papers have gauge length between these two values. The choice of gauge length often depends on the available mold for injection molding as well as the gauge length of the extensometer, which is mounted on the coupon for the measurement of loss of stiffness.

Horst *et al.* [HOR 97b] used plates that were 5.75 mm thick, while Jain *et al.* [JAI 16c] had coupons with a thickness of 2 mm. Thicker coupons usually lead to a skin core distribution of the FOD through the thickness of the coupon, thus making a direct correlation between the FOD and fatigue properties slightly difficult. In our view, it is advisable to use coupons with a gauge length as large as possible and a thickness as small as possible. Thinner samples are more likely to have uniform FOD through the thickness [VIN 05], making correlation between orientation and fatigue properties easier. Also, it is common practice to machine more than one coupon from one plate. However, it must be ensured that the FOD for coupons machined from different locations are the same since the properties of the coupons will depend on where the plate was machined. Edge clamps are rarely used for SFRC with coupons machined in a dog bone shape.

As was the case with textile composites (Chapter 1), the most common fatigue loading type is constant stress amplitude, sinusoidal wave-form of tension–tension loading with an applied load ratio of 0.1. The loading frequency typically varies from 1 to about 10 Hz, depending on the matrix, heat generated during loading and also the applied load.

Generally, the load and number of cycles to failure data is plotted in a log–log plot and the data are fit to linear plot, which is also called SN curve. Mathematically,

$$\sigma_{\text{fat}} = AN^{-1/k} \quad [3.1]$$

where σ_{fat} is applied maximum load, A is the intercept of the curve and slope of the SN curve is characterized by constant k. The two constants A and k are enough to characterize the curve.

The SN curve of SFRC depends on a number of factors that can be grouped as follows:

– fiber-based parameters including FOD, fiber length distribution (FLD) and fiber content;

– environmental conditions including temperature and moisture;

– specimen configurations including notches and weld lines;

– loading conditions that include applied load ratio (R-ratio) and multiaxial loading.

In the following sections, the SN curve dependence for each of the four subgroups is described in detail.

3.2.1. *Fiber-based parameters*

3.2.1.1. *Fiber orientation distribution (FOD)*

It is mostly believed that there is a strong dependence of fiber orientation on the fatigue behavior of SFRC. This has been confirmed by a number of studies [ARI 14a, DE 12, HAS 12, HOR 97b, JAI 16c, OGA 10] by testing coupons with different geometry, thickness or relative location of coupons in the mold. As described in the previous section, a common practice is injection to mold plates and machine coupons from them. This provides alternatives convenient to vary the FOD by machining coupons in different directions with respect to the MFD. It is usually not possible to create a direct relation between the second-order orientation tensor and fatigue behavior due to the skin core effect (leading to different second-order orientation tensor at different points through the thickness) and other deviations from perfect alignment, but it is generally believed that better alignment of fibers leads to higher strengths in that direction [LAP 08].

There is some disagreement about the slope of the SN curve. Guster *et al.* [GUS 11] claim that there is a strong dependency of the slope of the SN curve on the FOD, while Launay *et al.* [LAU 13] remarked that the slopes are independent of the FOD. The different slope in the SN curves, reported for certain experimental series, could be due to the scatter in the number of cycles to failure for a same applied load. Assuming a normal distribution of the cycles to failure for a given load, the statistical significance of the difference in slopes of two SN curves can be assessed using Student's *t*-tests [SCH 03].

In the following, the published fatigue data are collected for nine sets of fatigue data and used to confirm whether or not the slope of SN curve depends on the FOD (Table 3.1). The publication details of each paper are provided in the reference section so that the reader can access more details if interested. In each of the nine papers considered here, there were two or more SN curves reported for SFRC coupons with different FODs, but with the same fiber matrix composition. When there were more than two SN curves, the set of SN curves were compared by taking two curves at a time and repeating the calculations for every combination. For instance, if there were "p" number of SN curves for a certain material, the statistical comparison had to be repeated $p(p-1)/2$ number of times. All comparisons of the slopes are performed at a 95% significance level. The Student's *t*-test comparison of the slopes of the nine sets of published data is presented in Table 3.1.

It is seen that the difference in slope for most of the published data is small (as is evident from the small range of k-ratio for each published set) and the Student's *t*-test for variance confirms that the difference in the slope is not significant for all but one of the fatigue data available. The difference in the slopes was found to be significant for only one fatigue dataset (published by Bernasconi *et al.* [BER 07a]); it was found that while the 30°, 60° and 90° coupons had similar slopes, the slope of the 0° coupon was found to be statistically different to the other curves. It is thus concluded that it is reasonable to assume that the slopes of the SN curves are independent of the FOD, though there might be some exceptions.

Serial no.	Material	Source of data	*k*-ratio range	*T*-test result difference in slope
1	PBT-GF50	[JAI 16c]	11.9–14.6	Not significant
2	PA66-GF35	[DE 12]	15.4–17.2	Not significant
3	PA6-GF30	[BER 07a]	17.5–22.2	Significant
4	PA6-GF30	[HOR 96]	12.4–12.4	Not significant
5	PA6-GF33	[WYZ 04]	13.9–14.0	Not significant
6	PBT-GF30	[WYZ 04]	17.5–20.3	Not significant
7	PBT-GF30	[KLI 11a]	17.8–20.8	Not significant
8	PA66-GF30	[ARI 14a]	17.0–25.0	Not significant
9	PA66-GF33	[ZHO 06]	15.1–18.5	Not significant

Table 3.1. *Summary of published data and t-test result for comparison of slopes of SN curves. PBT, polybutylene terephthalate; PA, polyamide; GF, glass fiber; the number after GF indicates fiber weight fraction [JAI 16a]*

Another important consideration is the variation in the fatigue properties for the different grades of thermoplastic polymer which has been used. It is typical practice to name the thermoplastic used for making the composite rather than giving more details about the exact grade, its chemical composition and other additional constituents. A look at any material database will show that there are several varieties of a certain polymer with wide variation in the properties. For example, according to material database Campus plastics [PLA 14], there are over 500 different grades of commercially available polymer PA-6. There is a significant variation in the Young's modulus and yield stress properties of the different variants of poly-butylene-terephthalate (PBT).

Several authors [BER 07a, DE 12, HOR 97b, WYZ 05, WYZ 08] reported the SN curve(s) of glass fiber reinforced PA-66; the primary orientations of the fiber (with respect to the loading) are altered by machining test coupons in different directions with respect to the flow direction of the matrix. In each of the cases, the fiber content lies between 30 and 35% by weight. Depending on the primary direction of the FOD, the reported SN curves of the five papers are grouped in three categories: 0°, 90° and the others (30°, 60° and 45° coupons). The SN curves are superimposed in Figure 3.4. A clear dependence of strength on the orientation is seen; 90° coupons are weaker than the corresponding 0° coupons. However, it is also seen that for same matrix (but different grades), similar fiber content and orientation, there can be a large variation in the strength (up to a factor of 2–3) of SFRC. Different strengths can be explained by the fact that different grades of PA –66 could have varying degrees of adhesion and yield stress with the fibers and loss of strength in SFRC during cyclic loading can be largely attributed to fiber–matrix debonding.

A final note about the dependence on the FOD is regarding the variation of fatigue properties as a function of coupon thickness. A few authors tested the fatigue behavior of SFRC coupons with different thickness of coupons and reported that the behavior of the materials depended on the thickness of the coupon (for example [DE 12]). This dependence on the coupon thickness is actually due to different FODs. For the same injection molding conditions, the FOD variation through the thickness of the coupons is a function of the thickness [VIN 05]. Thicker coupons have a "skin core" variation of the FOD, while thinner coupons have more aligned FOD, thus there is greater anisotropy if thinner coupons are used for testing.

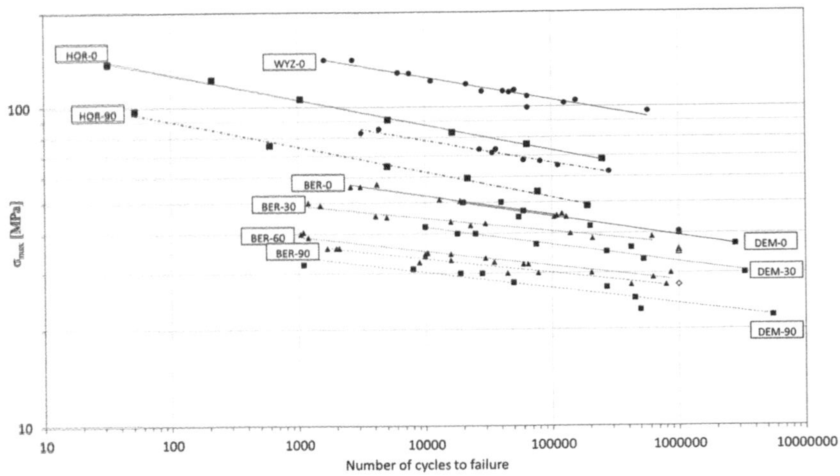

Figure 3.4. *Effect of the FOD on the SN curve of SFRC [BER 07a, DE 12, HOR 97b, WYZ 05, WYZ 08]. In all the cases shown in the figure, the material is PA –66 with glass fiber, fiber content in all the cases is between 30 and 35% weight fraction. Text box adjacent to the curves show the source of the paper followed by the orientation*

3.2.1.2. *Fiber length distribution (FLD)*

Compared to FOD, there has been less attention devoted to the FLD. Bernasconi *et al.* [BER 07b] pointed out that there is a direct relation between the average fiber length and fatigue strength of SFRC. However, the effect of FLD on fatigue properties is not straightforward to measure and quantify. This is because invariably longer fibers will lead to more scattered FOD. Two coupons that have different fiber lengths but exactly the same FOD are hard to fabricate, thus a direct comparison of the fatigue properties is not possible.

Meneghetti *et al.* [MEN 14] studied the fatigue behavior of SFRC with initial fiber lengths of 10 and 1 mm, after injection molding the average fiber length was found to be 3.4 and 0.72 mm, respectively. There is significant fiber breakage during the injection molding process as the fibers pass through the nozzle, screw, barrel, runners and gates of the injection molding setup. They surprisingly observed that the fatigue strengths of the SFRC with shorter fibers were better than the one with longer fibers. This was attributed to the more scattered FOD in the SFRC with longer fibers. The slope of the SN curves with different FLD was very similar to each other. Grove and Kim [GRO 95] performed fatigue tests on SFRC coupons with fiber lengths in the range of 0.16–0.28 mm and 1.18–1.37 mm. For axial loading, they observed that at low cycle regions, the coupons with longer fiber performed better with larger strengths. However, at a higher number of cycles the situation reversed and shorter fiber coupons performed better. This suggests that the slope of the two SN curves is different. Better fatigue strength at low cycle fatigue (LCF) suggests that at that region, fiber breakage is the primary mode of failure; while at high cycle fatigue (HCF) fiber matrix debonding becomes more important. For flexural loading, the longer fiber coupons performed better at both LCF and HCF.

Thus, it can be concluded that though it is more or less clear that the fatigue behavior of SFRC does indeed depend on the FLD, it is somewhat hard to obtain a clear trend.

3.2.1.3. *Fiber volume fraction*

As is the case with static strength, it is expected a decrease in the fatigue strength if the volume fraction of the fiber is reduced. Zago and Springer [ZAG 01] produced coupons made up of injection molded copolyamide matrix reinforced with 30 and 50% weight fraction of glass fiber and machined in 0° and 90° to the flow of the matrix direction. They reported that the fatigue strength of injection molded composites depended on the weight fraction of the fibers (Figure 3.5). The slope of SN curves for the 0° coupons was similar for the two volume fractions while the SN curves for the 90° coupons seemed to intersect.

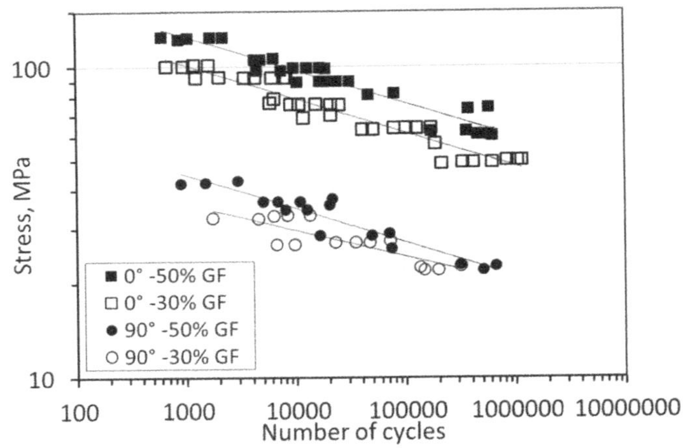

Figure 3.5. *Fatigue behavior of copolyamide reinforced with 30 and 50% weight fraction GF. Square markers indicate 0° coupons while circular markers indicate 90° coupons. Solid and hollow markers are indicative of 50 and 30% GF content, respectively. Data from [ZAG 01]*

Mandell *et al.* [MAN 83] studied the fatigue behavior of glass fiber reinforced polysulfone matrix subject to tension–tension load and observed that the *k*-ratio of the SN curve remained constant and close to 8 for SFRC with 10, 20, 30 and 40% glass fiber content. The strength of the composite increased as the fiber volume fraction increased. Interestingly, the *k*-ratio for 40% carbon fiber reinforced polysulfone matrix was also reported to be close to 8, while the *k*-ratio of unreinforced polysulfone matrix was only about 4.

Janzen and Ehrenstein [JAN 89] studied the fatigue behavior of glass fiber reinforced polybutylene terephthalate with a fiber volume fraction varying from 15 to 44%. They reported that fatigue strengths at low cycle regions were strongly dependent on the fiber content, with additional fibers leading to higher strength. However, at high cycle region, the strengths were found to be independent of the fiber content.

Other fiber-related factors include fiber surface treatment and sizing. Takahara *et al.* [TAK 94] and Yamashita *et al.* [YAM 98] showed an improvement in lifetime by about 15% with suitable treatment of the fiber surface. Surface treatment of the fiber led to both better adhesion as well as reduction of the stress concentration at the interphase. Both these factors lead to an improvement in the fatigue properties.

3.2.2. *Loading-based parameters*

3.2.2.1. *Frequency*

Esmaeillou *et al.* [ESM 13] showed that SN curves for tension–tension fatigue of glass fiber reinforced PBT at different frequencies (up to 20 Hz) are similar. However, they noticed a decrease in strength as a function of rising frequency for flexural fatigue tests. However, it is known that thermoplastics are generally poor conductors of heat and thus there is a lot of dissipative heating during testing at high frequencies, increasing the temperature of the coupon. This can lead to thermal fatigue.

To counter this, the frequency for fatigue tests is chosen on the basis of the temperature rise on the surface of the coupon [BEL 06, CAS 06, LAN 87a, LAN 87b]. A rise of about 5°C [DE 12] or 10°C [HOR 96] is deemed allowable for polyamides. Some authors have also used different frequencies for different loads (for example [HIT 93]); such an approach could be risky without controlling temperature rise.

3.2.2.2. *Mean stress*

Mallick *et al.* [MAL 04] and De Monte *et al.* [DEM 12] performed fatigue tests on SFRC with different mean stresses. Both have reported that the fatigue strength decreased with increasing mean stress. The effect of FOD seemed to decrease with increasing mean stress. This could be due to the reason that at high mean stress, the failure is mostly due to creep. Matrix is susceptible to creep while fiber is not. Oka *et al.* [OKA 07] performed fatigue tests for 30% glass fiber reinforced PBT, for seven different *R*-ratios. From their results, it was clear that the strengths tend to reduce as the applied load ratio increases, there also seem to be strong dependencies of the slope of the SN curve. The SN curves get flatter as the *R*-ratio increases (Figure 3.6).

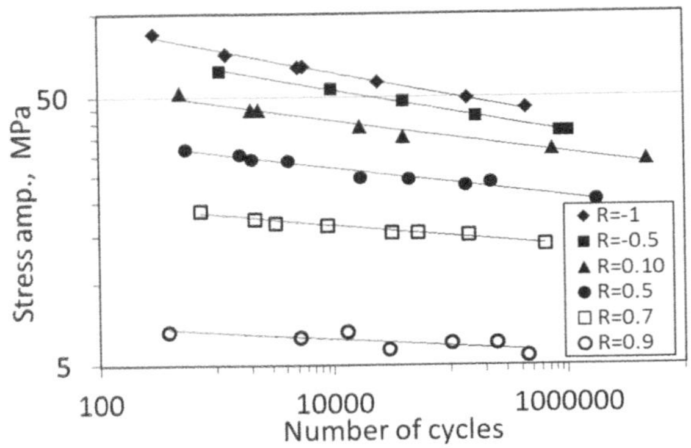

Figure 3.6. *Fatigue behavior of 30% glass fiber reinforced PBT for different R-ratio. Data from [OKA 07]*

3.2.2.3. *Multiaxial loads*

Compared to uniaxial tests, there have been very few publications discussing the behavior of SFRC when subjected to multiaxial fatigue (three examples are [DEM 09], [KLI 11b] and [SAW 16]). All of them subjected SFRC coupons to a combination of tensile and shear loading and confirmed that there are adverse effects on the life of SFRC when subjected to multiaxial loads and that shear loads are important for causing failure. It is clear from their data that the strength drops and that the slope of the SN curve becomes steeper when shear loading is combined with axial tension loading. It is not known whether or not the effect of multiaxial loads is independent of the FOD as shear fatigue with coupons with different FODs have not yet been reported.

De Monte *et al.* [DEM 09] used cylindrical test coupons and subjected these to a pure-tension, pure-torsion and mixed mode loading (tension–torsion). Two different mixed mode loads were applied by varying the multiaxial load parameter, λ, which was defined as the ratio of the shear and the tensile stress. They reported a drop in strength as the component of torsion is increased. However, the number of tests was too low to perform meaningful statistical analysis and confirm the slope dependence of the SN curve (Figure 3.7).

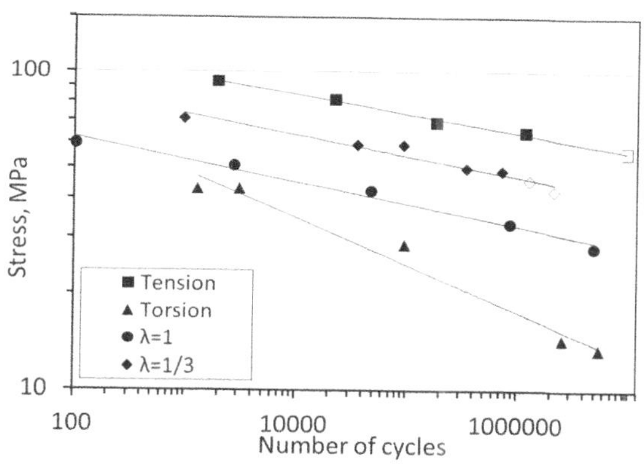

Figure 3.7. *Multiaxial fatigue behavior of 35% weight fraction GF-reinforced polyamide (PA) λ is the multiaxial load parameter; hollow markers indicate run outs. Data from [DEM 09]*

3.2.3. *Environmental effects*

Components made up of injection molded SFRCs are subject to different environments with different temperature and water content. Therefore, it is important to study the effects of moisture and temperature on fatigue behavior of SFRCs.

3.2.3.1. *Moisture and humidity*

Polyamide thermoplastics generally swell due to absorption of moisture, which leads to mismatched fiber and volume expansion.

Subsequently, internal stresses rise, which leads to an increase in the stresses in the interface and brings the stress closer to the interfacial strength of the fiber–matrix interface. This reduces the fatigue strength of SFRC [ARI 14b, BRA 95]. Barbouchi *et al.* [BAR 07] produced SN curves of the same material at different humidity levels and confirmed that there is a drop in strength as the levels of humidity increase, but the SN curves were shown to intersect, which points out that the slope of the SN curve is dependent of the moisture content (Figure 3.8). Hoppel [HOP 92] compared the fatigue behavior of dry short glass fiber reinforced styrene-maleic anhydride coupons and also coupons that were dipped in water for 2 days. They observed a drop in fatigue behavior both in terms of strength as well as quicker loss of stiffness.

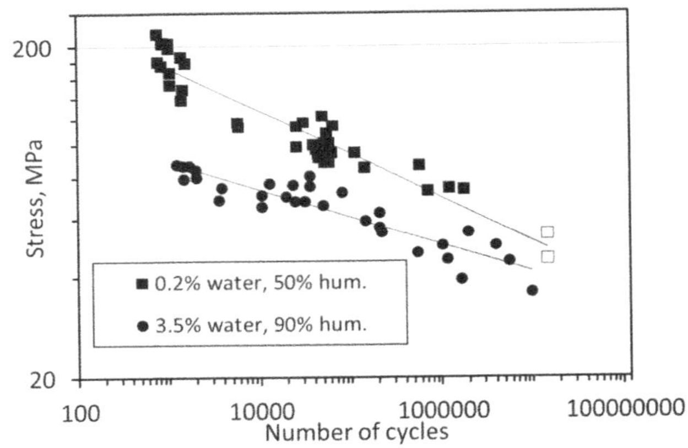

Figure 3.8. *SN curve for the 30% glass fiber reinforced polyamide –66, R = –1; f = 10 Hz, rhomboid markers indicate samples containing 0.2% H_2O tested at 23°C and 50% humidity, circular markers indicate samples containing 3.5% H_2O tested at 23°C and 90% humidity. Hollow markers indicate run outs. Data from [BAR 07]*

3.2.3.2. *Temperature*

There have been various attempts to study the fatigue behavior of SFRC at different temperatures [DE 12, DE 09, HAN 99, JIA 98, NOD 01]. It is generally agreed that the strengths decrease as the temperature is increased. However, the fatigue behavior of SFRC at extremely low temperatures is not known. Jia and Kagan [JIA 98] performed fatigue tests at –40, 23 and 121°C and observed that the highest fatigue strengths were observed at –40°C with significant drop in strength at 121°C. Handa *et al.* [HAN 99] argued that the dependence of intercept and slope of the SN curve on the temperature is linear up to the glass transition temperature; the slope and intercept decreases as the temperature increases. Above the glass transition temperature, the slope of the SN curve remains constant.

The slopes of the SN curve were found to be independent of temperature [DE 12]. There are no concrete results describing whether the dependence of fatigue behavior of temperature is the same for different FODs. According to De Monte *et al.* [DE 12], decreased fatigue strength at higher temperatures is largely due to the altered behavior of the matrix (and the interface), and thus the dependence of the fatigue behavior on temperature must be the same for different FODs.

Recall that the behavior of the SFRC often depends on the behavior of the matrix and the previously reported findings about dependency on the temperature and humidity might be completely different for a different matrix. Caution must be exercised even if one has the same polymer, since, as shown in section 3.1.1 (Figure 3.3), different grades for the same polymer can lead to significantly different behavior. This difference could become even more exaggerated when elevated temperatures and humidity levels are in question.

3.2.4. *Specimen configurations*

3.2.4.1. *Notch sensitivity and stress concentration effects*

Stress concentrations and notches are important considerations during the design of components subject to fatigue loading. This aspect becomes even more important due to the fact that the injection molding process facilitates the manufacturing of components with holes and irregular shapes. There have been a few studies conducted on the effect of geometric irregularities on the fatigue behavior of SFRC.

Zhou and Mallick [ZHO 06] studied the effect of circular hole with different diameters on the overall fatigue behavior. They tested glass fiber reinforced PP coupons with widths of 12.7 mm and holes with diameters of 1.58, 3.17 and 6.34 mm. The measured stress concentration factors were 2.64, 2.43 and 2.18, respectively. They reported that the presence of notches led to reduced fatigue performance. However, when the net surface area was considered, the SN curves for the three different holes collapsed into one. Haldar and Senthilvelan [HAL 11] showed that the effect of notch sensitivity is higher if the fiber length of the composite is higher. Two different sized notches and SFRC with two different fiber lengths (0.44 mm and 1.25 mm) were considered. At HCF, the effect of notch sized was significant for SFRC with longer fiber while, like Zhou and Mallick, they also reported that the fatigue behavior was independent of notch size for composites with smaller fibers. The longer fiber composites suffered from fiber pull outs leading to lowered fatigue behavior; this mode of damage was not present in the composite with shorter fibers.

Sonsino and Mossbrugger [SON 08] studied the combined effect of notches and loading at elevated temperatures. Using circular holes or slit notches, stress concentrations ranging from 1.0 to 9.8 were achieved. At room temperature, there was a direct relation between an increasing stress concentration factor and reducing fatigue performance. However, at 130°C the effect of the notches completely disappeared. This can be explained by the fact that at elevated temperatures the matrix becomes more ductile, which leads to reduced notch sensitivity.

All these studies were primarily focused on tension–tension loading (*R*-ratio = 0, 0.1). Mortazavian and Fatemi [MOR 14] reported the effect of notches on SFRC subjected to tension compression (*R*-ratio = –1.0) along with tension–tension fatigue. The effect of notches was seen to be more significant for compression loading, particularly at the HCF region.

There have been some sporadic efforts at understanding the effect of notches and it is commonly agreed that the fatigue behavior of SFRC does not depend on the diameter of the hole for tension–tension fatigue loading if the fibers are small in length. A systematic experimental campaign combined with thorough modeling to explain the findings would be very useful.

3.2.4.2. *Weld lines*

Apart from notches and stress concentrations that are functions of geometry of the component and are present in all components irrespective of the material. SFRC also have some specific manufacturing attributes that alter the fatigue behavior of the component. One important attribute is the weld line. When a liquid matrix fills up a mold from two opposing directions, a weld line is created. This weld line has a drastic effect on the fatigue properties of SFRC; Zhou and Mallick [ZHO 11] reported that the fatigue strength of SFRCs with weld lines was less than 50% in comparison to SFRCs with no weld line, with all other parameters kept constant.

To summarize, there are a large number of factors on which the SN curve of SFRC may depend. A description of the different factors including the SN curve behavior was provided in this section. In the following section, we look at another manifestation of the fatigue behavior, i.e. the loss of stiffness.

3.3. Loss of stiffness

Like most other polymer composite materials, SFRCs are also known to suffer loss of stiffness during cyclic loading. There is a large amount of experimental evidence that confirms the loss of stiffness [ARI 14a, ARI 14b, DE 12, KLI 11a, KLI 11b, MAL 04]. The extent of loss of stiffness seems to be dependent on the type of matrix and the degree of adhesion between the fiber and the matrix. De Monte *et al.* [DE 12] tested 35% glass fiber reinforced PA –66 and reported that there was a loss of about 10–15% stiffness, while Arif *et al.* [ARI 14a] tested 30% glass fiber reinforced PA –66 and reported that the loss of stiffness was only about 5% after million cycles of loading.

Very little is known about the dependence of the loss of stiffness on the various parameters described in section 3.1. We will now present a statistical analysis of the loss of stiffness during cyclic loading to demonstrate that the loss of stiffness is typically independent of FOD for same applied load. For the scope of this book, only tension–tension fatigue tests are considered, but the method developed in this book can be used for comparison of loss of stiffness curves at different applied load ratios as well [JAI 16b].

PBT reinforced with 0.5 weight fraction of glass fiber (equivalent volume fraction is 0.35) compound is injection molded to manufacture plates with dimensions of 170×170×2 mm^3. The thickness of the plates are chosen to be as thin as possible so as to ensure uniform FOD (with negligible core layer) through the thickness of the coupon based on the advice of Vincent *et al.* [VIN 05]. Coupons were machined from the plates in three directions, inclined at angles φ = 0°, 45° and 90° with respect to the prevailing flow direction (Figure 3.3(c)). For the rest of the book, they are referred to as 0°, 45° and 90° coupons. Three coupons were machined from each plate for the 0° and 90° coupon, while only one coupon per plate was machined for the 45° coupon. A schematic representation of the plate geometry, coupon dimensions and machining was presented in Figure 3.3.

Manufacturing simulation using software SIGMASOFT [SIG 14] shows that the FOD in the 0° coupon characterized by second-order tensor is:

$$a_{ij} = \begin{bmatrix} 0.81 & 0.018 & 0.137 \\ 0.018 & 0.11 & 0.004 \\ 0.137 & 0.004 & 0.079 \end{bmatrix} \qquad [3.2]$$

Physically, this means that a 0° coupon has more fibers that are oriented in the zero direction. The FOD for 45° and 90° coupons can be derived by simple rotation of the orientation tensor. It must be noted at this point that there are more accurate methods for characterizing the FOD, for example, using micro-CT techniques. However, for the analysis presented here, it is enough to simply confirm that the FOD for the 0°, 45° and 90° coupons are significantly different from each other.

The coupons are subject to tension–tension fatigue loading (R-ratio = 0.1); the applied load, in MPa, is calculated by dividing the applied load by cross-sectional area. The strain in the coupon was derived using extensometers that are mounted on the coupons. Coupons that survived a million cycles were counted as run outs and not used during the calculation of the SN curve. Run out coupons were subjected to quasi-static tensile loading until failure post the fatigue tests. The rate of loading for these tests was 0.5 mm/min. These tests provide a qualitative comparison between the pre- and postfatigue tensile behavior of SFRC. The results of the fatigue tests are shown in Figure 3.9. Expectedly, the coupon is significantly dependant on the fatigue strength of FOD. A student's *t*-test confirms that the slope of the SN curves is independent of the FOD.

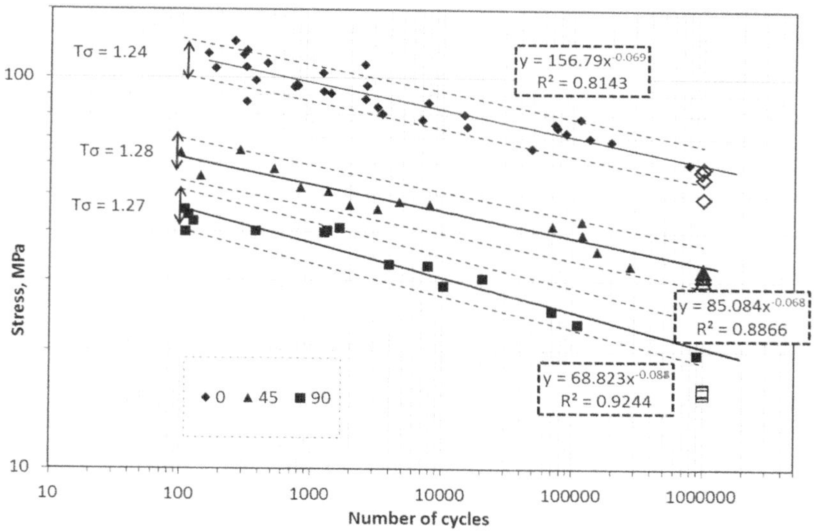

Figure 3.9. *SN curve for 50% wt. fraction glass fiber reinforced PBT. Applied load ratio is 0.1; hollow markers indicate run outs. Data from [JAI 16c]*

The pre- and postfatigue behavior of 0°, 45° and 90° coupons is presented in Figure 3.10. It is clear from the stress–strain curves that there is a significant loss of stiffness due to cyclic loading. However, the strain to failure seems to remain more or less the same. This probably hints that the damage mechanism due to cyclic loading is predominantly due to fiber matrix debonding while final failure during quasi-static loading is due to matrix failure. A summary of the measured properties is given below in Table 3.2.

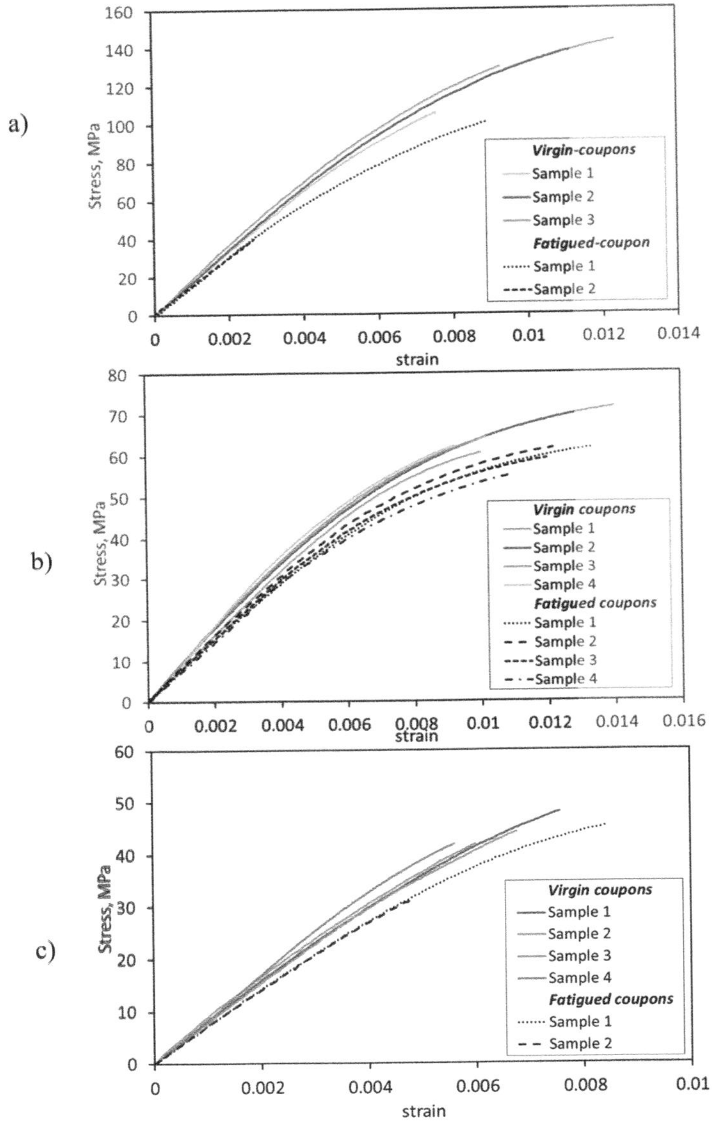

Figure 3.10. *Pre- and postfatigue behavior of 50% wt. fraction glass fiber reinforced PBT. Virgin coupons and coupons which survived million cycles are subject to quasi-static tensile loading: a) 0°, b) 45° and c) 90° coupons. Data from [JAI 16b]*

		0°	45°	90°
Virgin sample	Stiffness (GPa)	17.06 ± 0.47	9.15 ± 0.75	8.35 ± 0.465
	UTS (MPa)	126 ± 15	65 ± 4.8	44 ± 2.8
	Strain to failure (%)	0.97 ± 0.23	1.08 ± 0.21	0.63 ± 0.8
Fatigued sample	Stiffness (GPa)	14.9 ±-	8.2 ± 0.38	7.45 ±-
	UTS (MPa)	70.47 ±-	59.22 ± 3.04	38.2 ±-
	Strain to failure (%)	0.57 ±-	1.19 ± 0.08	0.66 ±-

Table 3.2. *A comparison of the stiffness, UTS and modulus of the virgin and fatigued samples for 0°, 45° and 90° coupons. The standard deviation for the fatigued 0 and 90° sample is not presented since the number of datasets is only 2. Data from [JAI 16b]*

3.3.1. *Collection of loss of stiffness data*

The stiffness of the coupon is tracked by recording the load in the fatigue machine and the displacement in the extensometer, which was mounted on the coupons. The data can be recorded for every cycle or for cycles with predetermined intervals. The secant stiffness of the coupon is then calculated by correlating the stress and strain in a linear fit. It was seen that the residual of the regression fit, R^2, was higher than 0.95 indicating the stiffness values are quite reliable. The stiffness is calculated for every cycle and counted as valid only if the value of the residual of the regression fit, R^2, is greater than 0.95. A particular fatigue experiment is counted as valid only if more than 95% of the recorded cycle data are found to be valid.

The SN curve (shown in Figure 3.9) is divided into three regions based on the number of cycles to failure. The division is done in such a way that there are at least three data points for each SN curve and the range of cycles per group is kept to a minimum. The range of number of cycles to failure for the three groups is as follows:

- Group 1 (high cycle fatigue region): 10^6 cycles;
- Group 2 (mid cycle fatigue region 279,940–20,673 cycles;
- Group 3 (low cycle fatigue region): 8,072–1,261 cycles.

The loss of stiffness curves for the three groups is shown in Figure 3.11.

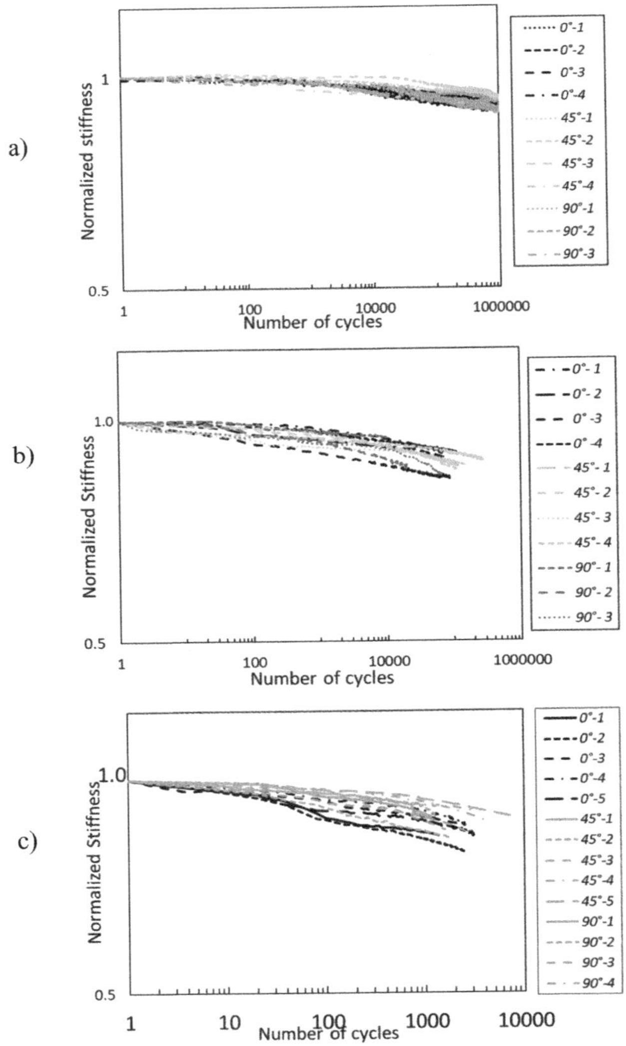

Figure 3.11. *Measured loss of stiffness curves for 0°, 45° and 90° coupons. Curves are grouped according to the number of cycles to failure: a) Group 1, b) Group 2 and c) Group 3. Data from [JAI 16b]*

3.3.2. *Comparison of the loss of stiffness curves*

Once the loss of stiffness curves has been successfully extracted, the next task is to develop a systematic method for comparing these losses. In this chapter, the loss of stiffness curves are fit to quadratic equations with intercept set to zero. This choice is based on the higher residual of regression values (as compared to linear fit) and due to the fact that the non-physical intercept of greater than one has been avoided.

Mathematically the quadratic fit equation is as follows:

$$\log{(E_N)} = \text{coeff1} \times \log N \times \log N + \text{coeff2} \times \log N \qquad [3.3]$$

In this equation, *coeff1* and *coeff2* are the two constants. Thus, each loss of stiffness curve is characterized by two constants, which will be compared by statistical tests. The chosen statistical test for this comparison is the Kruskal–Wallis test [KRU 52], commonly known as "one-way ANOVA on ranks". This test is the most reliable test for comparing small datasets (3–5 in our case) and datasets with unequal sample size. The p-value of the Kruskal–Wallis test on the data is calculated (with the help of in-built features in data processing software like Excel) and compared with the value of the chosen significance level. The hypothesis of similarity of data is supported if the p-value is higher than the value of alpha.

For the sake of comparison, two measured quantities will be compared: the first is the measured loss of stiffness and the second is the measured loss of stiffness curves.

Three steps are performed for a statistical comparison of the curves:

1) the measured loss of stiffness data is fit to the quadratic equation (equation [3.3]) and the value of R^2 is compared to ensure fit similarly to the power equation. The Kruskal–Wallis test can no longer be applied if there are large variations in the value of the residual of the regression fit;

2) the first coefficient of the curve is compared (equation [3.3]) and a Kruskal–Wallis test is performed;

3) the second constants are also compared in a similar manner.

The comparison of both coefficients is done for a confidence limit of 95%. Therefore, the hypothesis of the similarity of the different loss of stiffness curves is tested for a composite hypothesis of 90.25% (= 95 × 95%) significance; a 90% confidence interval is admittedly not a very stringent criterion. However, it is common practice to use a confidence interval of 90% for design analysis. A large number of fatigue design analysis of metals and composites using confidence intervals of 90% have been reported in the literature [CHE 05, DRA 14, GOP 12, KRU 52, LAS 05, REN 12].

For group 2 (mid cycle range), the measured loss of stiffness is shown in Figure 3.12(a). Similarly, the values of coefficients 1 and 2 are shown in Figures 3.12(b) and 3.12(c), respectively.

The value of p for the dataset presented for group 2 for measured loss of stiffness and coefficients 1 and 2 is 0.98, 0.70 and 0.46, respectively. Recall here that the variation in the properties can be considered insignificant if the p-value is higher than the value of the significance level (i.e. 0.05). Therefore, it can be confirmed that the loss of stiffness is independent of the FOD. A similar comparison for the low and high cycle region is presented in greater detail in [JAI 16b].

Loss of stiffness adds an additional degree of complexity to the simulation of fatigue. Evolution of stiffness and variation of loads and displacements occur at two completely different time scales. Loss of stiffness is slow and could occur after thousands of cycle but the load and displacement vary multiple times in one single cycle [FIS 11]. Therefore, fatigue simulation of SFRC is a temporal multiscale problem. This is in addition to the presence of different length scales in dimensions of constituents (fiber and matrix) and components.

a)

b)

c)

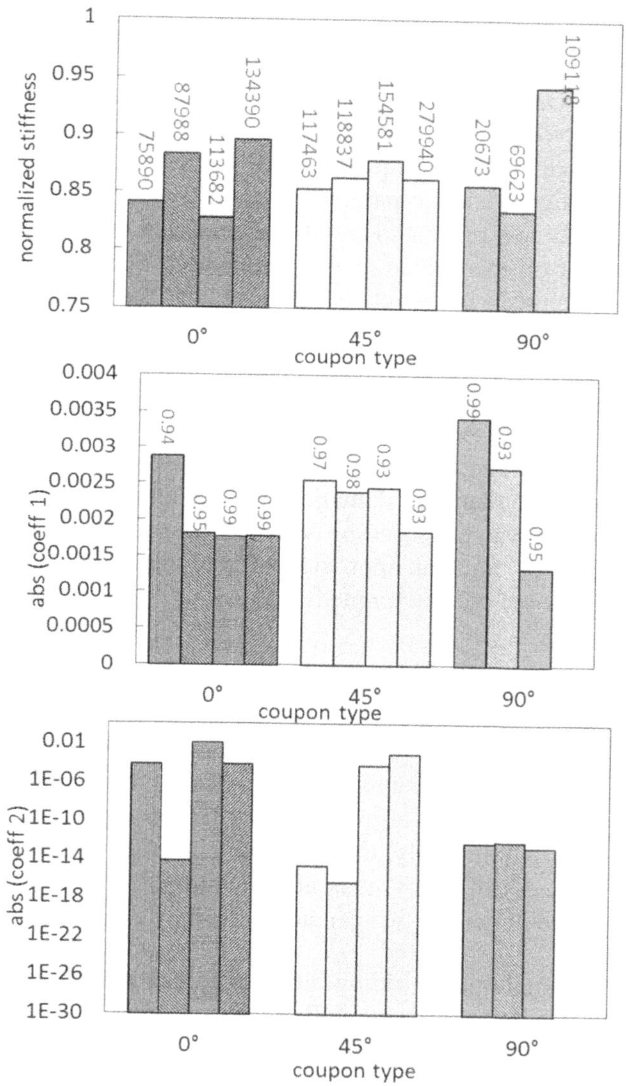

Figure 3.12. *Measured loss of stiffness quantities for 0°, 45° and 90° coupons that survived between 279,940 and 20,673 cycles before failure: a) normalized measured loss of stiffness, number of cycles to failure shown above bar; b) coefficient 1, the number over each bar is the value of R^2; c) coefficient 2. Data from [JAI 16b]*

There have been some attempts at modeling the loss of stiffness using constitutive equations, starting from the damage model of Ladaveze and Dantec [LAD 92], which deals with anisotropic damage of laminates. Nouri *et al.* [NOU 09] modeled the loss of stiffness for SFRC by modifying the equations of Ladaveze and Dantec and introducing fatigue damage parameters to deal with cyclic loading. A parameter study was presented and it was seen that the loss of stiffness depended a lot on the choice of the parameters. Such approaches are not well accepted in practice because of the extremely high computational times. As such, such an approach is deemed feasible only at the coupon level. Modeling loss of stiffness cycle by cycle is time consuming, and for a component level modeling, better strategies must be considered. One such strategy could be to model the loss of stiffness at cycles that are chosen on the basis of some predetermined criteria and interpolate the damage between those cycles. A "cycle jump" algorithm was proposed by Van Paepegem *et al.* [VAN 01] for textile composites. Such an approach could be interesting for SFRC as well but has not yet been attempted.

3.4. Future outlook and modeling strategy

By now the reader would have appreciated and noted the vast challenges for fatigue design and analysis of SFRC. The fatigue properties depend upon a large number of factors making the analysis quite complex. Additionally, there is also lack of understanding about several factors. As of today, it is almost impossible to build a single model or design a method that can accommodate all possible factors.

In such a situation, a building block approach toward the design of SFRC materials is recommended. Predictive models must first be built to account for individual factors (such as FOD, FLD and temperature). Efforts must then be made to integrate models that account for different factors. In some ways, this prescribed approach is similar to the "Chinese puzzle" approach [LOM 14]. It is clear that a simple linear superposition of the different factors is not appropriate; some advanced ideas on integrating different models must be developed.

A single model that accounts for all the factors will be hard to construct and even if it were to be built, the computational costs involved might be prohibitive in nature. Also, exact properties of the individual constituents (fiber, matrix and interphase) will be needed. This could be a potentially big bottleneck since the exact properties of the matrix will rarely be known to the designer. Generally, precompounded pellets with fiber and the matrix are supplied with only the name of the polymer disclosed by the material supplier. Models for predicting the fatigue behavior of SFRC as function of FOD, FLD and multiaxial loading will be presented in Chapter 4.

3.5. Bibliography

[ABD 14] ABDIN Y., LOMOV S.V., JAIN A. *et al.*, "Geometrical characterization and micro-structural modeling of short steel fiber composites", *Composites Part A: Applied Science and Manufacturing*, vol. 67, pp. 171–180, 2014.

[ADV 87] ADVANI S.G., TUCKER C.L. "The use of tensors to describe and predict fiber orientation in short fiber composites", *Journal of Rheology*, vol. 31, no. 8, pp. 751–784, 1987.

[ARI 14a] ARIF M.F., SAINTIER N., MERAGHNI F. *et al.* "Multiscale fatigue damage characterization in short glass fiber reinforced polyamide-66", *Composites Part B-Engineering*, vol. 61, pp. 55–65, 2014.

[ARI 14b] ARIF M.F., MERAGHNI F., CHEMISKY Y., "In-situ damage mechanisms investigation of PA –66/gf –30 composite: effect of relative humidity", *Composites Part B-Engineering*, vol. 58, pp. 487–495, 2014.

[BRA 95] BRADLEY W.L., GRANT T.S., "The effect of the moisture absorption on the interfacial strength of polymeric matrix composites", *Journal of Material Science*, vol. 30, pp. 5537–5542, 1995.

[BAR 07] BARBOUCHI S., BELLENGER V., TCHARKHTCHI A. *et al.*, "Effect of water on the fatigue behaviour of a PA –66/glass fibers composite material", *Journal of Materials Science*, vol. 42, pp. 2181–2188, 2007.

[BEL 06] BELLENGER V., TCHARKHTCHI A., CASTAING P., "Thermal and mechanical fatigue of a PA –66/glass fibers composite material", *International Journal of Fatigue*, vol. 28, pp. 1348–1352, 2006.

[BER 07a] BERNASCONI A., DAVOLI P., BASILE A. *et al.*, "Effect of fibre orientation on the fatigue behaviour of a short glass fibre reinforced polyamide-6", *International Journal of Fatigue*, vol. 29, no. 2, pp. 199–208, 2007.

[BER 07b] BERNASCONI A., DAVOLI P., ROSSIN D. *et al.*, "Effect of reprocessing on the fatigue strength of a fibreglass reinforced polyamide", *Composites Part A-Applied Science and Manufacturing*, vol. 38, no. 3, pp. 710–718, 2007.

[BER 08] BERNASCONI A., COSMI F., DREOSSI D., "Local anisotropy analysis of injection moulded fibre reinforced polymer composites", *Composites Science and Technology*, vol. 68, pp. 2574–2581, 2008.

[BIR 12] BIRON M., *Thermoplastics and Thermoplastic Composites*, Elsevier Science, 2012.

[BRA 95] BRADLEY W.L., GRANT T.S., "The effect of the moisture absorption on the interfacial strength of polymeric matrix composites", *Journal of Materials Science*, vol. 30, pp. 5537–5542, 1995.

[CAS 06] CASADO J. A., CARRASCAL I., POLANCO J.A. *et al.* "Fatigue failure of short glass fibre reinforced PA –66 structural pieces for railway track fasteners", *Engineering Failure Analysis*, vol. 13, pp. 182–197, 2006.

[CHA 00] CHANG S., HWANG J., DOONG J., "Optimization of the injection molding process of short glass fiber reinforced polycarbonate composites using grey relational analysis", *Journal of Materials Processing Technology*, vol. 97, pp. 186–193, 2000.

[CHE 05] CHEN Z., ZHENG S., "Lifetime distribution based degradation analysis", *IEEE Transactions on Reliability*, vol. 54, pp. 3–10, 2005.

[DEM 09] DE MONTE M., MOOSBRUGGER E., JASCHEK K. *et al.* "Multiaxial fatigue of a short glass fibre reinforced polyamide 6.6: fatigue and fracture behaviour", *International Journal of Fatigue*, vol. 32, pp. 17–28, 2009.

[DEM 12] DE MONTE M., MOOSBRUGGER E., QUARESIMIN M., "Influence of temperature and thickness on the off-axis behaviour of short glass fibre reinforced polyamide 6.6 cyclic loading", *Composites Part A: Applied Science and Manufacturing*, vol. 41, no. 10, pp. 1368–1379, 2012.

[DRA 14] DRAPER N.R., SMITH H., *Applied Regression Analysis*, John Wiley & Sons, 2014.

[EIK 13] EIK M., LOHMUS K., TIGASSON M. *et al.*, "DC-Conductivity testing combined with photometry for measuring fibre orientations in SFRC", *Journal of Materials Science*, vol. 48, pp. 3745–3759, 2013.

[ESM 13] ESMAEILLOU B., FERREIRA P., BELLENGER V. *et al.*, "Fatigue behavior of polyamide 66/glass fiber under various kinds of applied load", *Polymer Composites*, vol. 33, pp. 540–547, 2013.

[FIS 11] FISH J., "Multiscale modeling and simulation of composite materials and structures", in DE BORST R., RAMM E. (eds), *Multiscale Methods in Computational Mechanics*, Springer, pp. 215–231, 2011.

[GOP 12] GOPE P., "Scatter analysis of fatigue life and prediction of s–n curve", *Journal of Failure Analysis and Prevention*, vol. 12, pp. 507–517, 2012.

[GRO 95] GROVE D., KIM H., Fatigue behavior of long and short glass reinforced thermoplastics, SAE Technical Paper, paper no. 950561, 1995.

[GUI 93] GUILD F.J., SUMMERSCALES J., "Microstructural image-analysis applied to fiber-composite materials – a review", *Composites*, vol. 24, pp. 383–393, 1993.

[GUS 11] GUSTER C., PINTER G., MOSENBACHER A. *et al.*, "Evaluation of a simulation process for fatigue life calculation of short fibre reinforced plastic components", *11th International Conference on the Mechanical Behavior of Materials (ICM11)*, Milan, Italy, 2011.

[HAL 11] HALDAR A., SENTHILVELAN S. "Notch effect on discontinuous fiber reinforced thermoplastic composites", *Key Engineering Materials*, vols. 471–472, pp. 173–178, 2011.

[HAN 99] HANDA K., KATO A., NARISAWA I., "Fatigue characteristics of a glass-fiber-reinforced polyamide", *Journal of Applied Polymer Science*, vol. 72, pp. 1783–1793, 1999.

[HAS 12] HASHIMOTO M., OKABE T., SASAYAMA T. *et al.*, "Prediction of tensile strength of discontinuous carbon fiber/polypropylene composite with fiber orientation distribution", *Composites Part A-Applied Science and Manufacturing*, vol. 43, pp. 1791–1799, 2012.

[HIT 93] HITCHEN S.A., OGIN S.L., "Damage accumulation during the fatigue of an injection moulded glass/nylon composite", *Composites Science and Technology*, vol. 47, pp. 83–89, 1993.

[HOP 92] HOPPEL C.P., "The effects of tension-tension fatigue on the mechanical behavior of short fiber reinforced thermoplastics", *Proceedings American Society for Composites*, Switzerland, pp. 509, 1992.

[HOR 96] HORST J.J., SPOORMAKER J.L., "Mechanisms of fatigue in short glass fiber reinforced polyamide 6", *Polymer Engineering and Science*, vol. 36, no. 22, pp. 2718–2726, 1996.

[HOR 97a] HORST J.J., *Influence of Fibre Orientation on Fatigue of Short Glassfibre Reinforced Polyamide*, TU Delft, The Netherlands, 1997.

[HOR 97b] HORST J.J., SPOORMAKER J.L., "Fatigue fracture mechanisms and fractography of short-glass fibre-reinforced polyamide 6", *Journal of Materials Science*, vol. 32, no. 14, pp. 3641–3651, 1997.

[JAI 16a] JAIN A., VAN PAEPEGEM W., VERPOEST I. *et al.*, "A feasibility study of the master SN curve approach for short fiber reinforced composites", *International Journal of Fatigue*, vol. 91, pp. 264–274, 2016.

[JAI 16b] JAIN A., VAN PAEPEGEM W., VERPOEST I. *et al.*, "A statistical treatment of the loss of stiffness during cyclic loading for short fiber reinforced injection molded composites", *Composites Part B: Engineering*, vol. 103, pp. 40–50, 2016.

[JAI 16c] JAIN A., VEAS J.M., STRAESSER S. *et al.*, "The Master SN curve approach – a hybrid multiscale fatigue simulation of short fiber reinforced composites", *Composites Part A: Applied Science and Manufacturing*, vol. 91 no. 2, pp. 510–518, 2016.

[JAN 89] JANZEN W., EHRENSTEIN G.W. "Hysteresis measurements for characterizing the cyclic strain and stress sensitivity of glass fiber reinforced PBT", *ANTEC'89: Plastics Create a World of Difference*, pp. 659–662, 1989.

[JIA 98] JIA N., KAGAN V.A., "Effects of time and temperature on the tension-tension fatigue behavior of short fiber reinforced polyamides", *Polymer Composites*, vol. 19, pp. 408–414, 1998.

[KAR 89] KARGERKOCSIS J., FRIEDRICH K., "Skin-core morphology and humidity effects on the fatigue crack-propagation of pa-6.6", *Plastics and Rubber Processing and Applications*, vol. 12, pp. 63–68, 1989.

[KIM 01] KIM E.G., PARK J.K., JO S.H. "A study on fiber orientation during the injection molding of fiber-reinforced polymeric composites: comparison between image processing results and numerical simulation", *Journal of Materials Processing Technology*, vol. 111, pp. 225–232, 2001.

[KLI 11a] KLIMKEIT B., NADOT Y., CASTAGNET S. *et al.* "Multiaxial fatigue life assessment for reinforced polymers", *International Journal of Fatigue*, vol. 33, no. 6, pp. 766–780, 2011.

[KLI 11b] KLIMKEIT B., CASTAGNET S., NADOT Y. *et al.*, "Fatigue damage mechanisms in short fiber reinforced PBT+PET GF30", *Materials Science and Engineering: A*, vol. 528, pp. 1577–1588, 2011.

[KRU 52] KRUSKAL W. H., WALLIS W.A., "Use of ranks in one-criterion variance analysis", *Journal of the American Statistical Association*, vol. 47, pp. 583–621, 1952.

[LAD 92] LADEVEZE P., LEDANTEC E., "Damage modelling of the elementary ply for laminated composites", *Composites Science and Technology*, vol. 43, pp. 257–267, 1992.

[LAN 87a] LANG R.W., MANSON J.A., "Crack tip heating in short-fiber composites under fatigue loading conditions", *Journal of Materials Science*, vol. 22, pp. 3576–3580, 1987.

[LAN 87b] LANG R.W., MANSON J.A., HERTZBERG R.W., "Mechanisms of fatigue fracture in short glass fiber-reinforced polymers", *Journal of Materials Science*, vol. 22, pp. 4015–4030, 1987.

[LAS 05] LASSEN T., DARCIS P.H., RECHO N., "Fatigue behavior of welded joints part 1—statistical methods for fatigue life prediction", *Welding Journal*, vol. 84, pp. 183–187, 2005.

[LAS 08] LASPALAS M., CRESPO C., JIMENEZ M.A. *et al.*, "Application of micromechanical models for elasticity and failure to short fibre reinforced composites – numerical implementation and experimental validation", *Computers & Structures*, vol. 86, pp. 977–987, 2008.

[LAU 13] LAUNAY A., MAITOURNAM M.H., MARCO Y. *et al.*, "Multiaxial fatigue models for short glass fiber reinforced polyamide – part I: nonlinear anisotropic constitutive behavior for cyclic response", *International Journal of Fatigue*, vol. 47, pp. 382–389, 2013.

[LOM 14] LOMOV S.V., VERPOEST I., "Chinese puzzles of textile composites properties", *Proceedings of the 16th European Conference on Composite Materials (ECCM-16)*, Seville, Spain, 2014.

[MAL 04] MALLICK P.K., ZHOU Y., "Effect of mean stress on the stress-controlled fatigue of a short e-glass fiber reinforced polyamide-6,6", *International Journal of Fatigue*, vol. 26, pp. 941–946, 2004.

[MAN 83] MANDELL J.F., MCGARRY F.J., HUANG D.D. *et al.*, "Some effects of matrix and interface properties on the fatigue of short fiber-reinforced thermoplastics", *Polymer Composites*, vol. 4, pp. 32–39, 1983.

[MAR 85] MARSHALL I. H., BARNARD P. M., BUTLER R.J. *et al.*, "Fatigue scatter of ud glass epoxy, a fact or fiction?", *Composite Structures 3*, Springer, The Netherlands, pp. 69–82, 1985.

[MEN 14] MENEGHETTI G., RICOTTA M., LUCCHETTA G. *et al.*, "An hysteresis energy-based synthesis of fully reversed axial fatigue behaviour of different polypropylene composites", *Composites Part B: Engineering*, vol. 65, pp. 17–25, 2014.

[MOR 14] MORTAZAVIAN S., FATEMI A., "Notch effects on fatigue behavior of thermoplastics", *Advanced Materials Research*, vol. 891, pp. 1403–1409, 2014.

[NOD 01] NODA K., TAKAHARA A., KAJIYAMA T., "Fatigue failure mechanisms of short glass-fiber reinforced nylon 66 based on nonlinear dynamic viscoelastic measurement", *Polymer*, vol. 42, pp. 5803–5811, 2001.

[NOU 09] NOURI H., MERAGHNI F., LORY P., "Fatigue damage model for injection-molded short glass fibre reinforced thermoplastics", *International Journal of Fatigue*, vol. 31, pp. 934–942, 2009.

[OGA 10] O'GARA, J.F., NOVAK G.E., WYZGOSKI M.G., "Predicting the tensile strength of short glass fiber reinforced injection molded plastics", *Proceedings of the 10th-Annual SPE® Automotive Composites Conference & Exhibition (ACCE)*, Troy, MI, 2010.

[OKA 07] OKA H., NARITA R., AKINIWA Y. *et al.*, "Effect of mean stress on fatigue strength of short glass fiber reinforced polybuthyleneterephthalate", *Key Engineering Materials*, vol. 340, pp. 537–542, 2007.

[PLA 14] PLASTICS, CAMPUS, "CAMPUS® – a material information system for the plastics industry", available at http://www.campusplastics.com/, 2014.

[REN 12] REN X.D., ZHANG Y.K., JIANG D.W. *et al.*, "A model for reliability and confidence level in fatigue statistical calculation", *Theoretical and Applied Fracture Mechanics*, vol. 59, pp. 29–33, 2012.

[SAS 13] SASAYAMA T., OKABE T., AOYAGI Y. *et al.*, "Prediction of failure properties of injection-molded short glass fiber-reinforced polyamide 6,6", *Composites Part A-Applied Science and Manufacturing*, vol. 52, pp. 45–54, 2013.

[SAW 16] SAWADA T., AOYAMA H., "Effect of molding processes on multiaxial fatigue strength in short fibre reinforced polymer", *Fracture and Structural Integrity*, vol. 38, pp. 92–98, 2016.

[SCH 03] SCHNEIDER C.R.A., MADDOX S.J., Best practice guide on statistical analysis of fatigue data, The Welding Institute Statistics Report, 2003.

[SIG 14] SIGMA, "SIGMA Engineering GmbH, Aachen", available at http://www. sigmasoft.de/, 2014.

[SON 08] SONSINO C.M., MOOSBRUGGER E., "Fatigue design of highly loaded short-glass-fibre reinforced polyamide parts in engine compartments", *International Journal of Fatigue*, vol. 30, pp. 1279–1288, 2008.

[SUT 00] SUTHERLAND H.J., VEERS P.S., "The development of confidence limits for fatigue strength data", *2000 ASME Wind Energy Symposium*, pp. 413–423, 2000.

[TAK 94] TAKAHARA A., MAGOME T., KAJIYAMA T., "Effect of glass fiber-matrix polymer interaction on fatigue characteristics of short glass fiber-reinforced poly (butylene terephthalate) based on dynamic viscoelastic measurement during the fatigue process", *Journal of Polymer Science Part B: Polymer Physics*, vol. 32, pp. 839–849, 1994.

[TEM 90] TEMPLETON P.A., "Strength predictions of injection molding compounds", *Journal of Reinforced Plastics and Composites*, vol. 9, pp. 210–225, 1990.

[THO 02] THOMASON J.L., "The influence of fibre length and concentration on the properties of glass fibre reinforced polypropylene: 5. Injection moulded long and short fibre PP", *Composites Part A: Applied Science and Manufacturing*, vol. 33, pp. 1641–1652, 2002.

[TOD 94] TODD R.H., ALLEN D.K., ALTING L., *Manufacturing Processes Reference Guide*, Industrial Press, Inc., Norwalk, 1994.

[UNI 16] UNIVERSITY OF ALBERTA INDUSTRIAL DESIGN, "Brockey 'Injection moulding' licensed under creative commons attribution 3.0 via Wikimedia commons", University of Alberta Industrial Design, available at http://commons.wikimedia.org/wiki/File:Injection_m, 2016.

[VAN 01] VAN PAEPEGEM W., DEGRIECK J., DE BAETS P., "Finite element approach for modelling fatigue damage in fibre-reinforced composite materials", *Composites Part B-Engineering*, vol. 32, pp. 575–588, 2001.

[VIN 05] VINCENT M., GIROUD T., CLARKE A. *et al.*, "Description and modeling of fiber orientation in injection molding of fiber reinforced thermoplastics", *Polymer*, vol. 46, pp. 6719–6725, 2005.

[WYZ 04] WYZGOSKI M.G., KROHN J.A., NOVAK G.E., "Fatigue of fiber-reinforced injection molded plastics. I: stress-lifetime data", *Polymer Composites*, vol. 25, no. 5, pp. 489–498, 2004.

[WYZ 05] WYZGOSKI M.G., NOVAK G.E., "Predicting fatigue S-N (stress-number of cycles to fail) behavior of reinforced plastics using fracture mechanics theory", *Journal of Materials Science*, vol. 40, no. 2, pp. 295–308, 2005.

[YAM 98] YAMASHITA A., HIGASHI S., KOMATSU S.T. *et al.*, "Fatigue analysis of short glass fibre-reinforced nylon 6 on the basis of dynamic viscoelastic measurements under cyclic fatigue", *Key Engineering Materials* vol. CXXXVII, pp. 147–154, 1998.

[WYZ 08] WYZGOSKI M.G., NOVAK G.E., "An improved model for predicting fatigue s-n (stress-number of cycles to fail) behavior of glass fiber reinforced plastics", *Journal of Materials Science*, vol. 43, pp. 2879–2888, 2008.

[ZAG 01] ZAGO A., SPRINGER G.S., "Constant amplitude fatigue of short glass and carbon fiber reinforced thermoplastics", *Journal of Reinforced Plastics and Composites*, vol. 20, pp. 564–595, 2001.

[ZHO 06] ZHOU Y.X., MALLICK P.K., "Fatigue performance of an injection-molded short e-glass fiber-reinforced polyamide 6,6. i. effects of orientation, holes, and weld line", *Polymer Composites*, vol. 27, pp. 230–237, 2006.

[ZHO 11] ZHOU Y., MALLICK P.K., "Fatigue performance of injection-molded short e-glass fiber reinforced polyamide-6, 6. II: effects of melt temperature and hold pressure", *Polymer Composites*, vol. 32, pp. 268–276, 2011.

[ZHO 13] ZHOU H., *Computer Modeling for Injection Molding: Simulation, Optimization, and Control*, Wiley, 2013.

4

Fatigue Modeling of SFRC: A Master SN Curve Approach

In this chapter, a methodology for the fatigue simulation of short fiber reinforced composite (SFRCs) is presented. First, an overview of the fatigue simulation strategy starting from process simulation all the way up to component level fatigue simulation will be presented. Four steps will be identified for the fatigue simulation; one by one each of these four steps will be covered.

4.1. Overall framework and modeling strategy

Problems in fatigue, like many other engineering problems, have two different methods that could be employed to approach the problem. The first is that of a scientist who is interested in understanding reality in the highest possible details and developing models to describe his understanding. The second route is that of an engineer who needs tools to safely design parts that are subject to complex load histories [QUA 10]. These two approaches, which are usually contrasting in nature, have been balanced in this book. While the overall goal is to describe a model for fatigue simulation of SFRC; the understanding of fatigue and damage mechanisms are also given due importance.

One usually employs a *micro–macro* multiscale approach for fatigue simulation of SFRC. In this approach for component-level simulation, each element in the finite element (FE) model is a representative volume element (RVE), where each RVE has a known statistical distribution of orientation of fibers. Micromechanical analysis is performed at the RVE level and macroscopic analysis is performed at the component-level simulation. The stiffness and fatigue properties (to be calculated by microscale analysis) are needed at every point for the macroscopic analysis.

Commercially available software such as MoldFlow [AUT 14] or academic codes can be used for the simulation of the manufacturing process and for predicting the local fiber orientation distribution (FOD). There are also methods and even commercial software such as Converse [CON 14] and DIGIMAT [DIG 12] to predict the local stiffness based on the local FOD determined by the manufacturing simulation software. Next, the finite elements mesh of the component during the manufacturing simulation is mapped to FE software so that the correct local stiffness can be attributed to each element. Similar to the local stiffness, some model must be developed to derive the local SN curve as a function of the FOD so that SN curve based component-level simulation can be performed. A schematic representation of the process is shown in Figure 4.1.

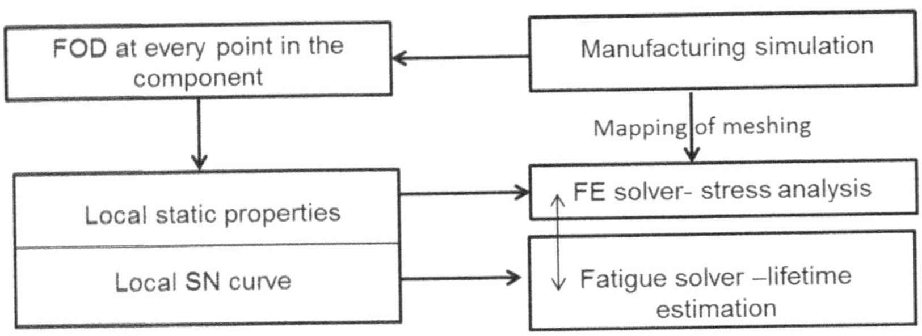

Figure 4.1. *A simplistic schematic representation of fatigue simulation process of SFRC*

According to Horst *et al.* [HOR 97], the calculation of the orientation distribution is the first step in order to determine the fatigue performance of an injection molded part. However, to be able to perform a fatigue simulation on a component we must be able to relate the tests on coupons to the fatigue behavior of a component. For a complete simulation of fatigue in SFRC, it is necessary to calculate the mechanical and fatigue behavior at every point on the component. There are a number on methods to calculate the effective stiffness of SFRC (for example [MOR 73]). However, there are no analytical methods available to estimate the fatigue behavior of composite materials and obtain their SN curve. There have been very few systematic attempts to build models to relate the SN curve to the microstructure of SFRC.

It is tempting to use empirical reduction factors for different factors such as FOD, length, notch and mean stress. Such an approach is often done for polymers [ERH 99]. A similar approach was tried by Guster *et al.* [GUS 11] and Mosenbacher *et al.* [MOS 14]. Such an approach will require an extremely large number of tests and can also be criticized for lacking a physical basis.

One of the more popular approaches used recently is the method in which the fatigue strength data are normalized by the ultimate tensile strength (UTS). This approach is based on experiments that showed the proportionality of both properties [BER 07a, DEM 12, BER 07b, KLI 13, ZAG 01]. A modified Tsai–Hill criterion is seen as a natural extension of this experimental fact [DEM 12, SPA 14]. Coefficients can be identified experimentally, but then the coefficients will be identified for one specific skin–core ratio and the modified Tsai–Hill criterion cannot be used on industrial structures where both thickness and fiber alignment degree are variable. Such an approach has two major limitations, first there is no physical reasoning as to why the fatigue strengths must vary in the same manner as the tensile strength. There is experimental evidence that confirm the failure mechanisms for damage propagation and final failure during static and fatigue loading are different [ARI 14a, ZAG 01, HOR 97]. Also to be able to use this method one has to either model tensile strengths of SFRC or experimentally derive the UTS for every orientation. To the best of our knowledge, there are yet no reliable methods to predict the tensile strength of SFRC accurately.

Wyzgoski *et al.* [WYZ 04] investigated a method to estimate the fatigue performance quickly from single cycle or quasi-static strength data based on an "empirical law" for polyamide. This empirical law indicated that the normalized fatigue strength decreased by 10% per point in the logarithmic scale of the number of cycles to failure. However, some deficiencies were detected when comparing the predictions with the results of the tests of different materials. While for nylon this approach represented a conservative estimation of fatigue lifetime, for other materials the empirical law prediction leads to incorrect predictions.

Some studies based on a fracture mechanics approach have been proposed and tried with varying degrees of success [KLI 11, KLI 13, WYZ 05]. Apart from a SN curve, the input for such an approach is a fatigue crack growth parameter. Experimental tests to derive the fatigue crack growth parameter are quite complex.

Damage leads to energy dissipation, and energy dissipation leads to increase in heat generation. Based on this principle, Meneghetti *et al.* [MEN 10] and Jegou *et al.* [JEG 13] proposed a method for SN curve prediction of SFRC based on the specific heat dissipation. Such an approach requires rather extensive and difficult tests.

A simple solution to this problem could be to experimentally extract the SN curve for extreme orientations and derive the SN curves at every point by a variation of interpolation [VER 13]. Such an approach has limited physical basis and requires a large amount of test data. Furthermore, it is almost impossible to create test coupons with perfect extreme alignment; therefore, some extrapolation might be needed as well.

There exists also a family of methods that aim to relate the fatigue properties of the composite with the fatigue properties of the fiber and the matrix [REI 91, ABD 15a, ABD 15b]. Such an approach has not been commonly used for SFRC since the fatigue properties of the matrix are not commonly available. This is because SFRCs are produced by injecting precompounded glass fiber reinforced thermoplastics, which makes testing on the pure matrix impossible.

Once the SN curves are derived, multiaxial fatigue can be treated using some fatigue strength criteria. Such an approach was advocated for composite materials by using strength criteria by [BER 07a, DEM 12]. They displayed good match of the fatigue strength with a Tsai–Hill type criteria. Other failure criteria that could be used are principal stress, Manson–Coffin equation, principal strain, modified energy criteria, etc. An in-depth description of the different multiaxial fatigue criteria can be found in [PAP 97, DEG 01]. Using the data of Klimkeit *et al.* [KLI 11] and De Monte *et al.* [DEM 12], Launay *et al.* [LAU 13] evaluated the different failure criteria for composite materials and confirmed that the modified energy criterion was the best failure criteria for fatigue strength for SFRC. This approach is valid only on a coupon-level simulation as the local orientation in a component varies from point to point [LAU 13].

In this chapter, we present a method for estimating the local SN curves by linking microscale damage to macroscale properties. This hybrid multiscale approach, which is named the master SN curve (MSNC) approach, relates the damage at the microscopic level to the macroscopic fatigue properties. The key innovative idea in this process is that it combines test results and simulation results (hybrid approach) on different scales (microscopic simulation, macroscopic fatigue behavior), overcoming problems of depending on too many expensive tests (pure test-based macroscopic approach) and lack of efficiency (pure microscopic simulation approach).

There are four key steps toward the development of the MSNC approach and subsequent component level SN curve.

First, it is not known whether the prevalent approaches to mean field homogenization are applicable for modeling damage and fatigue. Modeling damage in SFRC requires the knowledge of stresses in individual inclusions and an appropriate homogenization method, which has superior predictive ability for stresses in individual inclusions must be ascertained.

SFRC exhibits a nonlinear stress–strain behavior when subjected to tensile loading. The nonlinear stress–strain behavior is due to matrix nonlinearity and complex damage modes such as fiber–matrix debonding and matrix cracking. Each damage mode and its interaction with other damage modes must be modeled correctly to account for the nonlinearity. This is the second step toward development of fatigue simulation framework.

The third and probably the most critical step is to link the damage at the microscale (fiber, matrix and the interphase) to the macroscopic fatigue properties. This relation will be used to calculate the local SN curve as a function of the FOD.

Besides the scientific challenges described above, one also encounters challenges of linking several software to achieve a proper fatigue simulation of SFRC. Fatigue simulation of SFRC depends on co-ordination between manufacturing simulation software, a multiscale platform to calculate the effective properties, FE software to calculate the macro-stress and a fatigue solver. Other challenges for fatigue simulation include modeling fatigue behavior for multiaxial loads and accounting for loss of stiffness during cyclic loading. In general, the existing methods for fatigue simulation of composites assume ideal constant amplitude uniaxial loading. Due of the fact that SFRC components are subject to complex multiaxial loads, methods to deal with complex load scenarios must be developed. All these scientific and logistic challenges are grouped together and described in section 3.5.

A simplistic schematic representation of the fatigue simulation process is given below in Figure 4.2.

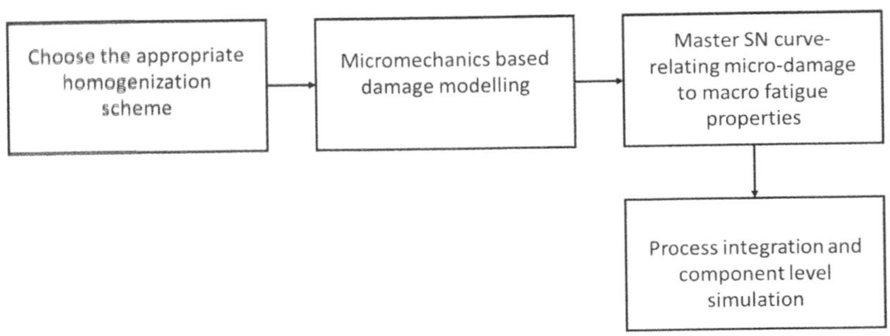

Figure 4.2. *Logical sequence of model building for fatigue simulation of SFRC*

Each of the four steps is described in detail in the following sections. Attempts to confirm the appropriate mean field homogenization scheme will be described in section 4.2. Next, the treatment of fiber matrix debonding, which is critical during fatigue simulation, will be elaborated upon in section 4.3. Section 4.4 is devoted to linking the damage at the interphase to the macroscopic fatigue behavior and development of the MSNC approach, which is a hybrid multiscale method to calculate the local fatigue properties as a function of the FOD. Finally, attempts at component-level simulation will be described in section 4.5. Section 4.6 presents the future outlook.

4.2. Choice of a mean field homogenization method

The effective properties of SFRCs can be estimated by a number of methods. Mean field homogenization methods based on the solution of Eshelby [ESH 57] are a family of methods, which aim to estimate the effective properties of a composite by relating it to the mean strains in the constituent phases. A number of mean field homogenization schemes exist, for example the Mori–Tanaka (MT) method [MOR 73], self-consistent scheme [HIL 65] and double interpolative inclusion [LIE 99]. The most common among them for the simulation of SFRC is the MT formulation.

The MT formulation is often criticized for giving physically inadmissible solutions in some specific cases. It is widely believed that the MT formulation is suitable for composites with reinforcements of similar shape and orientation (biphase composites) or if the distribution of orientations is statistically homogeneous random. However, the MT formulation is often used for multiphase composites. In the context of the inclusion theory, multiphase composites are composites that consist of inclusions with a statistical distribution of length and orientation. It is this extension from two-phase composites to multiphase composites and not the original MT assumption that was said to occasionally produce physically inadmissible results.

To circumvent the mathematical problems of the MT formulation, a pseudo-grain discretized approach to MT formulation was proposed first by Camacho *et al.* [CAM 90], extended by Doghri and co-workers [DOG 03, PIE 04, DOG 05] for a wide range of material behavior that was later incorporated to commercial industrial software DIGIMAT [DIG 12] and Converse [CON 14]. A pseudo-grain was defined as a biphase composite consisting of inclusions having the same orientation and aspect ratio (AR). They applied the MT formulation individually on the "pseudo-grains", and then volume averaged the stiffness of the grains to get the effective properties of the short-fiber composite. A schematic representation of the proposed scheme is recalled in Figure 4.3 [PIE 04].

The basic idea behind breaking the homogenization scheme into two steps is the following: if each step individually satisfies all the conditions of the homogenization scheme, then the procedure in itself will satisfy all the conditions required for mean-field homogenization schemes. This approach eliminated the mathematical problems of the MT formulation but introduced additional approximations with regard to the interactions between the inclusions. This Pseudo-Grain discretized implementation of the Mori-Tanaka formulation (PGMT) approach has been very popular for use in design and simulation of industrial components.

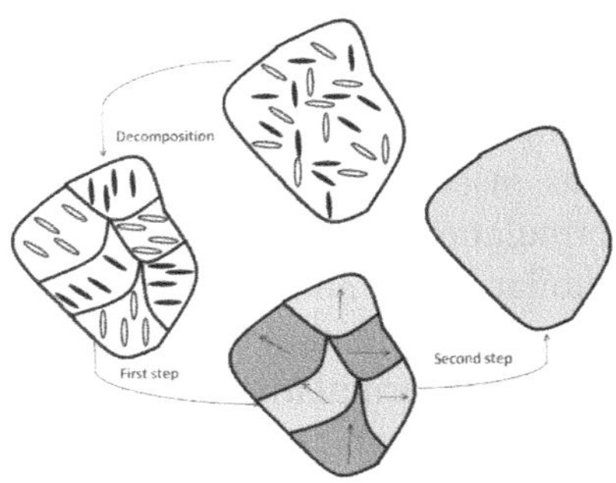

Figure 4.3. *Schematic representation of the pseudo-grain discretization. A multi-phase composite is discretized to multiple "pseudo-grains" and each pseudo-grain is homogenized separately in the second step to estimate the effective properties [PIE 04]*

Both the "full" MT and PGMT have been tested for their capabilities for predicting the effective stiffness [DOG 05, GHO 12, GUS 97,

KAR 07, TUC 99]. However, the modeling of damage requires more than just correct and reliable predictions of effective properties. Indeed, it can be said the correct prediction of effective stiffness is only the first step toward modeling of damage; micromechanical modeling of damage necessities that the microstresses (stresses in individual fibers and the matrix) be correctly predicted. Average stresses in individual inclusions are required to model damage events such as fiber–matrix debonding and fiber failure [HUY 01], while the average stresses in the matrix are usually used to model the material nonlinearity in the matrix. For the scope of fatigue modeling, it can be said that the correct prediction of the average stresses in individual inclusions is as important as the prediction of effective properties.

To ascertain the predictive abilities of the mean field homogenization schemes, solution of carefully built full FE models can be treated as quasi-correct and used as benchmark. Full FE calculations are computationally expensive rendering them unsuitable for multiscale analysis. However, such models present possibilities for benchmarking and comparing different predictive schemes. An example of such a comparison is described below.

4.2.1. *Benchmarking of schemes with full FE solution*

Let us consider an RVE that has a two-dimensional (2D) random distribution of inclusions, each with AR of 3 and the volume fraction of the inclusions is 0.25. A microstructure of the RVE can be generated using simple algorithms such as random sequential adsorption algorithm [RIN 97]. For the sake of easy meshing, the minimum allowed distance between two inclusions was 0.035 times the AR of the inclusion [PIE 04]. This minimum distance ensures a good quality mesh and subsequently reliable FE solution. Also, the minimum distance between the surface of the ellipsoid and the prism faces, edges or corners was ensured to be larger than 0.05 times the AR to avoid distorted elements during meshing. The ABAQUS solver was used to solve the FE problem with C3D10M elements – a 10 node tetrahedron element; this was as prescribed by the DIGIMAT software. An example of FE generated microstructure including meshing is shown in Figure 4.4.

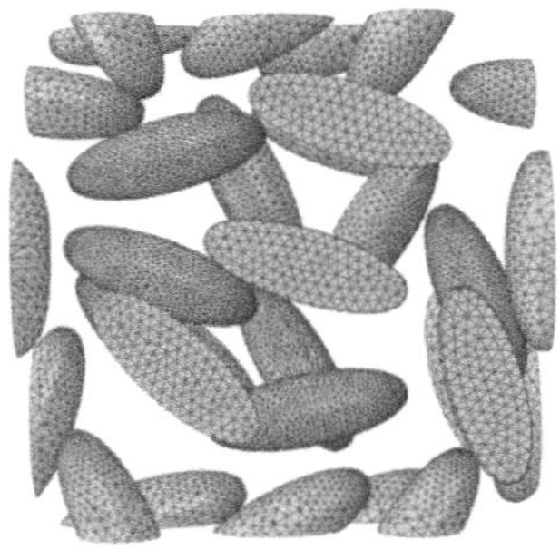

Figure 4.4. *Periodic cuboid showing FE RVE. Notice that the structure is periodic and inclusions intersecting a face of the cube also appear on the opposite face. Shown in the figure is VE containing 30 inclusions having an approximately uniform random distribution of inclusions with orientation tensor (a11 = 0.52, a22 = 0.48) [JAI 13]*

Once the FE model has been built, the next step is to ensure the representativeness of the RVE that has been built. In simple terms, it must be ensured that the results thus obtained by the full FE calculations would not be too different if a larger number of inclusions were considered. To ensure this, a number of FE models with a varying number of inclusions can be created and the minimum number of inclusions, which lead to maximum 5% variation in the average stresses is considered for further model building. For the configuration described in Figure 4.4, eight dispersions of VE with 5, 10, 20 and 30 inclusions were created. A uniaxial strain of 1% in the direction X is applied to each and the phase average stresses for each VE are calculated. For 20 inclusions, the difference between the maximum and minimum value of average stresses in the VE was less than the desired 5%, the maximum and minimum values of phase average stresses in the different realizations of the VE was 173.6 and 165.8 MPa, respectively. Thus, it was concluded that 20 or more inclusions would be needed to characterize a VE.

After creating the FE–VE, a full MT scheme is implemented with 1,000 inclusions and PGMT scheme is implemented by discretizing the RVE to 30 grains. Three models, viz. full FE, MT, and PGMT, are all subject to periodic boundary conditions and applied load of 0.1%. The volume fraction of inclusions is 0.1 for all the three cases.

The average stresses in inclusions are a direct output of the two mean field homogenization scheme, whereas the FE calculations yield the stresses in each element. The average stresses in individual inclusions for the FE model is calculated by volume weighted averaging of the stresses of the elements constituting the inclusion.

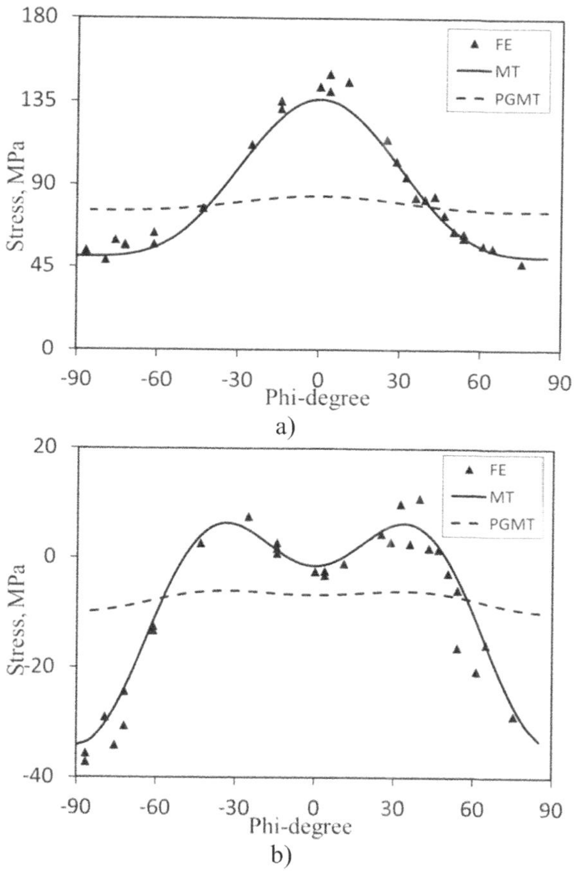

Figure 4.5. *Inclusion average stresses for approximation of uniform random orientation of inclusions, applied load is 1% strain in axial direction, vf = 0.1, with orientation tensor (a11 = 0.51, a22 = 0.49). a) Stresses in the global loading direction, S11. b) Stresses in the transverse direction, S22 [JAI 13]*

A comparison of the predicted average stress in the individual inclusion for the three cases is shown above in Figure 4.5.

It can be seen that the full MT scheme has superior predictive abilities of stresses in individual inclusions as compared to PGMT. In the MT formulation, interactions between the inclusions are taken into account by introducing the image strain. Each inclusion feels the presence of other inclusions indirectly through the total strain of the matrix. The additional strain on the inclusions due to the presence of other inclusions is called image strain. In the MT formulation, the image strain is equally distributed among all the inclusions. Thus, all the inclusions feel the presence of each other equally.

However, in the PGMT formulation, the inclusions are first separated and grouped into different pseudo-grains. The MT formulation is applied only at the pseudo-grain level, thus an inclusion receives image strain only from inclusions having the same AR and orientation. Therefore, the interactions between dissimilar inclusions are lost; this is the reason why the stresses in individual inclusions are predicted incorrectly. There are different values of the image strain per pseudo-grain as opposed to uniform image strain in the MT formulation. The PGMT formulation separates inclusions at a scale which it should not; this leads to loss of some interactions between the inclusions and also introduces spurious interactions between inclusions having the same orientation and AR.

However, both the schemes lead to very similar values of the effective properties. Thus, we conclude that either of the two variations of the MT scheme can be used if only the effective stiffness of the RVE is relevant. However, for damage and fatigue modeling where the average stresses in individual inclusions are important, a full MT formulation is advocated. An in depth comparison of the MT and PGMT schemes including different length ratios, AR and mixed fiber lengths can be found in [JAI 13]. It is acknowledged that the MT formulation can yield an asymmetric stiffness tensor, which is physically inadmissible, but the degree of asymmetry was shown to be rather low when the formulation was used to homogenize SFRCs [JAI 15b]. All the calculations presented henceforth in this book have been performed using the full MT calculations.

4.3. Damage modeling

SFRC suffer from a number of damage modes such as fiber–matrix debonding, matrix damage and fiber breakage. A number of authors have used different optical techniques for damage characterization of injection molded glass fiber composites. One of the earlier attempts was by Horst and Spoormaker [HOR 97], who used a fractography technique to study the fractured surface after fatigue and tensile failure and confirmed the presence of fiber–matrix debonding at the tip of inclusions, matrix cracking and fiber breakages. Cosmi *et al.* [COS 13, COS 11], Bernasconi *et al.* [BER 10a] and Arif *et al.* [ARI 14b] used Micro-CT techniques to study the evolution of damage events during tensile and fatigue loading. Different events observed were fiber–matrix debonding, fiber pull out and matrix cracking (Figure 4.6).

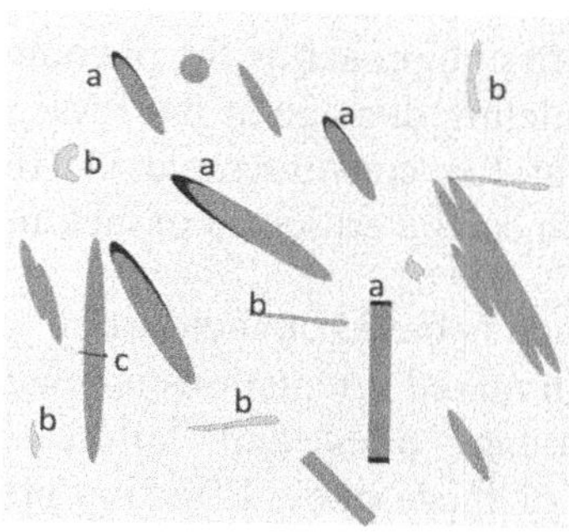

Figure 4.6. *Simplistic representation of different reported damage mechanisms observed by micro-CT in tensile specimen of SFRCs: a) fiber-matrix debonding; b) matrix micro cracks; c) fiber breakages*

There are a number of challenges that one encounters during micromechanical modeling of SFRC. First, the orientation distribution of fibers must be correctly predicted and characterized. Several experimental techniques such as measurements of fiber shapes in cross-sections, X-ray radiography or tomography, electron microscopy, confocal laser scanning microscopy, optical diffraction and reflected-light microscopy have been used to study the FOD of SFRC [DEM 10, EBE 01, FAR 04, HIN 95, MIE 12, MLE 99]. It can also be calculated using manufacturing simulation software [ZHO 13]. The FOD is usually characterized by a second-order orientation tensor [ADV 87], and efficient methods are available for constructing an RVE from the second-order orientation tensor [CIN 95, JAC 04, ONA 88].

Once an RVE has been created, the next challenge is to calculate both the effective properties of the composite and locally the stresses in individual fibers under a given loading. For this purpose, a number of mean field homogenization techniques have been developed, the most common among these is the MT formulation [MOR 73]. It was shown to predict in a reasonable manner both the effective properties as well as the stresses in individual inclusions (section 4.2). In the MT formulation, fibers are treated as ellipsoidal inclusions, thus the terms "fiber" and "inclusion" are often used interchangeably. The next step after the MT formulation is to model damage in SFRC.

Damage modeling approaches in SFRC can be broadly grouped in three categories. The first approach is the phenomenological approach which consists of modeling damage at the RVE level [ZHO 05]. The details and damage at the constituent level (fiber and matrix) are ignored. This approach can be criticized of lacking physical basis.

The second approach is based on the idea of "ply failure". An RVE made up of fibers with mixed orientations is imagined to be composed of several unidirectional plies and failure criteria are applied individually on each of those plies. Effective properties are typically calculated by averaging the stiffness across different plies [HAL 78, KAR 85]. This idea of damage modeling is highly approximate since it is based on the damage modeling of a ply, which in reality does not exist in an SFRC. Such an approach cannot fully take into account the interactions between fibers with different orientations.

The third approach consists of modeling damage at the constituent level [MER 95]. Nonlinearity in SFRC could be caused by fiber–matrix debonding, fiber breakage or matrix nonlinearity; these damage mechanisms are confirmed by a number of experimental studies [ARI 14a, GUP 90, HOR 96, HOR 97, MOU 06, SAT 91]. These damage events are accounted for during micromechanics-based damage modeling of SFRC. This micromechanics based damage model is the preferred approach to modeling damage since actual damage modes are being modeled.

In this book, we advocate the use of micromechanics-based damage modeling with special emphasis on fiber–matrix debonding. A schematic representation of the proposed scheme is shown in Figure 4.7.

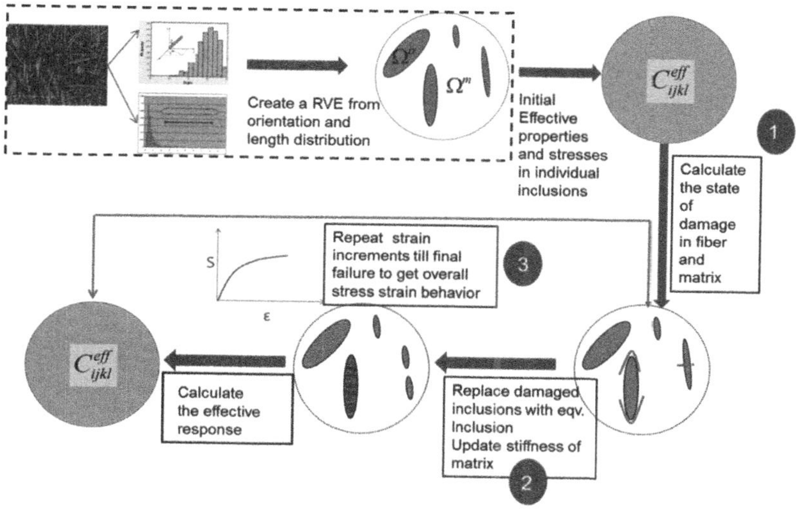

Figure 4.7. *Three steps for modeling the overall stress–strain behavior of SFRC, images top left corner are from [JAO 05]*

4.3.1. *Fiber–matrix debonding: equivalent bonded inclusion approach*

Fiber–matrix debonding refers to the damage in the interface between the fiber and the matrix. The properties of a SFRC depend on the properties of its constituents and on the capacity of stress transfer in the interface [THO 02]. The treatment of fiber matrix debonding has two important components: first the onset of debonding must be calculated; next the mean field homogenization method must be adapted for accommodating the inclusions with imperfect interphase.

Within the framework of the MT formulation, the most convenient method for calculating the onset of debonding is the use of the modified Coulomb's criteria. This criterion takes into account both shear as well as normal stress. Mathematically, the condition for fiber–matrix debonding is

$$\sigma_N + \beta\tau > \sigma_{crit} \qquad [4.1]$$

where σ_N and τ are the normal and tangential stress component at the interface, β is the shear contribution coefficient and σ_{crit} is the strength of the interface. This formulation can be used to model debonding at every point of the inclusion, even though the MT formulation only yields the average stress of the inclusion. This is because at every point on the inclusion surface, the normal and tangential stresses are calculated and onset of debonding is calculated as a function of those stresses (equation [4.1]). A detailed description of the relationships between the average stress in the inclusion and the interface was provided by Huysmans *et al.* [HUY 01].

The shear contribution coefficient, β, is an empirical quantity and cannot be directly measured by simple experiments. The usual procedure is to assume a given value in the range of 0.4–0.5 [HUY 01, JAO 05]. However, appropriate value of β can also be ascertained by parametric studies and choosing the value of β, which leads to best fit to the experimental stress–strain curve.

The use of the modified Coulomb's criteria at every point in the surface of the fiber leads to prediction of exact extent and location of the region where there is imperfect interphase. The debonded interface cannot be treated directly within the framework of the mean field homogenization since the mean field homogenization formulations are based on the assumption of perfect adhesion between the fiber and the matrix. In this book, we propose to replace the debonded inclusion (fiber) by a fictitious "equivalent bonded inclusion" (EqBI) with modified stiffness, so that the Eshelby [ESH 57] solution and consequently the mean field homogenization schemes could still be used.

A fiber with debonded interface is replaced by an EqBI. The stiffness properties of the EqBI are calculated in such a way that the EqBI is mechanically equivalent to the debonded inclusion. To estimate the stiffness tensor of an EqBI, a volume element (VE) consisting of a single inclusion is assumed and approximate expressions for the stress (re)distribution in the inclusion due to the debonded interface are derived. The average stresses in the inclusions are then calculated. The diagonal terms, C'_{ii}, of the stiffness tensor of the EqBI are determined as a product of (1) the corresponding diagonal terms (C_{ii}) of the stiffness tensor of the original inclusion and (2) the ratio of the average stress in the inclusion with debonded interface to the average stress that would exist in the inclusion, if it was perfectly bonded:

$$C'_{ii} = \frac{<\sigma'_{ii}>}{<\sigma_{ii}>} C_{ii} \qquad [4.2]$$

where C'_{ii}, C_{ii} are the stiffness component of the EqBI and the original inclusion that relates the average stress σii in a VE to the applied strain ε_{ii}, σ'_{ii} and σ_{ii} are the stresses in the inclusion with debonded and perfect interface, respectively; < > indicates volume averaging. The non-diagonal terms of the stiffness tensor are calculated assuming that the Poisson's ratio of inclusion does not change. By this method, it can be ensured that the effective stress response of the two systems, viz. RVE with EqBI, and the original RVE with fiber having debonded interface is the same. A schematic representation of the idea is shown in Figure 4.8.

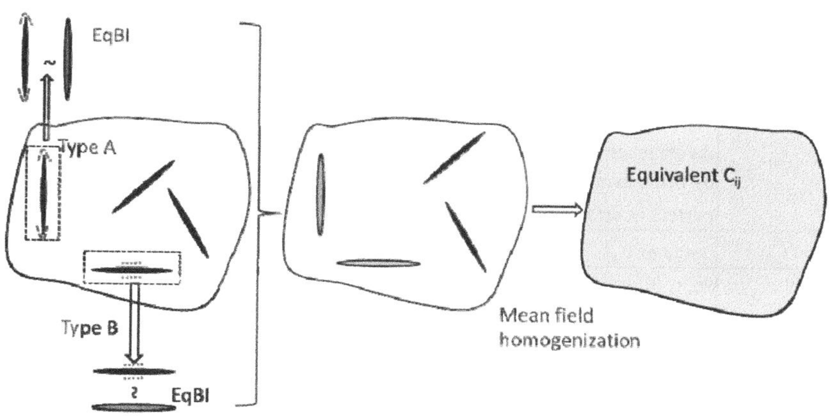

Figure 4.8. *Schematic representation of the concept of the equivalent bonded inclusion (EqBI) used to calculate the effective properties of SFRC containing inclusions with debonded interface. Two different configurations of the debonded interface (type A and B) [JAI 15a]*

Based on the location of the debonded region, the type of debonding can be grouped to two categories: Type A-tip debonding and Type B-debonding at equator. Equations can be derived for stress redistribution due to the presence of debonded interphase for both Type A and B debonding for the six possible cases of uniaxial loading; they have been summarized in Table 4.1. For an in-depth description of the proposed scheme including validation of the mechanical equivalence, the reader is advised to refer [JAI 15a].

At the end of each time step, the damaged fiber is replaced by EqBI and/or broken fibers depending on the predicted damaged state and subsequently the MT calculations as well further damage checks are performed on the modified RVE, which contains undamaged inclusions as well as EqBI and/or broken fibers (Figure 4.7).

<table>
<tr><th>Stiffness component</th><th>Type A</th><th>Type B</th></tr>
<tr><td>C_{zz}</td><td>$\sigma'_{incl} = E_f.\varepsilon_1 - (E_f.\varepsilon_1 - 2.\tau_i.s.m).\frac{\cosh\left(\frac{nz}{r}\right)}{\cosh(n.s(1-m))}$
where, $\tau_i = -\mu(\sigma_{res} - \nu_1.E_m.\varepsilon_1)$ and $n^2 = \frac{2.G_m}{E_f.\ln\frac{R}{r}}$
Average stress is calculated by integration over length of inclusion, z; other terms have usual meanings</td><td>No change in average stress</td></tr>
<tr><td>C_{yy}</td><td colspan="2">$\sigma'_{incl} = \sigma_f$ if interface is perfect
$\sigma'_{incl} = 0$ if interface is debonded
Average stress is calculated by volume weighted averaging
σ_f is the average stress in the inclusion that would be present if there was perfect interface in the inclusion</td></tr>
<tr><td>C_{xx}</td><td colspan="2">Same as C_{yy}</td></tr>
<tr><td>C_{xy}, C_{xz}, C_{yz}</td><td colspan="2">$\sigma'_{incl} = \sigma_f$ if interface is perfect
$\sigma'_{incl} = -\mu\sigma_{rr}$ if interface is debonded and σ_{rr} is negative
$\sigma'_{incl} = 0$ if interface is debonded and σ_{rr} is positive
Average stress is calculated by surface area weighted averaging of σ'_{incl}
σ_{rr} is the stress component in the outward normal direction of the inclusion if there was perfect interface in the inclusion</td></tr>
</table>

Table 4.1. *A summary of the expressions of the stresses in inclusions with debonded interface, σ'_{incl}. The stiffness of the EqBI is calculated as a function of the reduced stress in the inclusion with the debonded interface [JAI 15a]*

The input for the modeling of onset and progression of fiber matrix debonding is the elastic properties of fiber and the matrix (modulus and Poisson's ratio), yield stress of the matrix and the configuration of the SFRC material (fiber volume fraction, length and FOD).

4.3.2. *Matrix damage*

Matrix nonlinearity is implemented using the well-known secant approach originally proposed by Tandon and Weng [TAN 88] and implemented for SFRC by Jao Jules *et al.* [JAO 05]. The input for such an analysis is the stress–strain curve of the matrix. For a given applied strain to the RVE, the average strain in the matrix is derived using the MT formulation. For this average strain in the matrix, the corresponding stress in the matrix is extracted from the input stress–strain curve of the matrix and the modulus of the matrix is calculated. Next, the MT formulation is repeated to calculate the stiffness of the composite and to recalculate the average strain in the matrix to confirm convergence. The above steps are iterated till convergence is achieved.

The simulations are stopped and the SFRC is assumed to have failed if the von- Mises strain in the matrix is higher than the strain corresponding to the highest stress of the matrix. The input for modeling matrix damage by this model is the stress strain curve of neat resin.

Due to the relatively short length of the fibers, fiber breakage does not occur during loading and hence is not considered during the simulations.

4.3.3. *Validation*

The proposed models described in sections 4.3.1 and 4.3.2 are validated and shown in Figure 4.9. The stress strain data for 50% weight glass fiber reinforced polybutylene terephthalate (PBT) is used for the validation (the coupon preparation was described in Figure 3.3 and the SN curve was shown in Figure 3.6). Three sets of simulation performed for each type of coupon (considering only fiber–matrix debonding, matrix nonlinearity and both together) are presented.

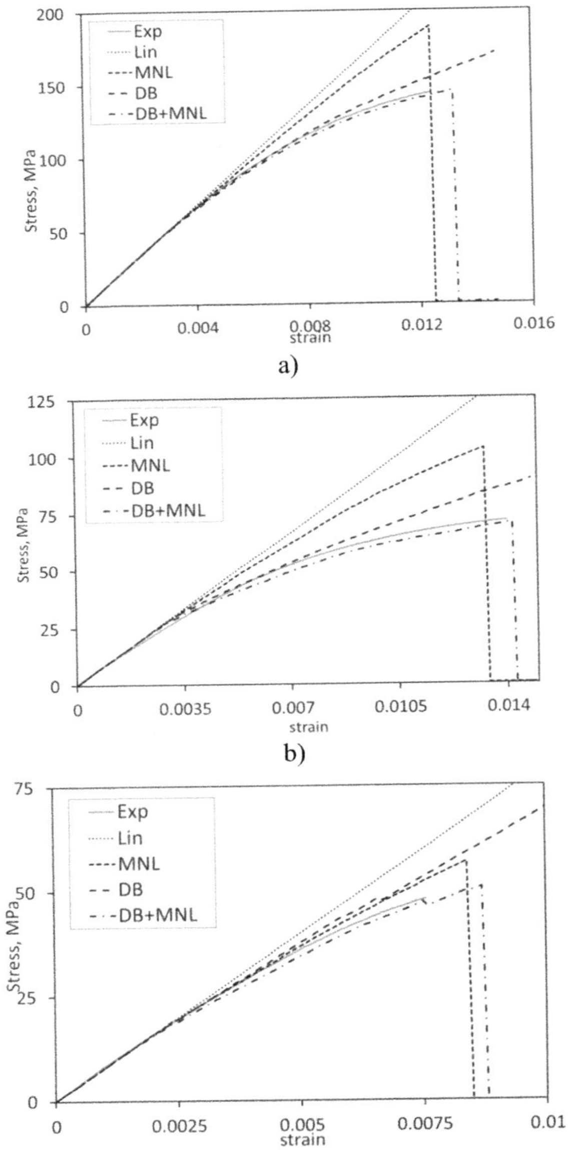

Figure 4.9. *Simulation of stress–strain behavior by considering linear behavior (Lin), matrix nonlinearity only (MNL), fiber–matrix debonding only (DB) and the both together (DB+MNL): a) 0°; b) 45°; c) 90° coupons*

It is seen that an accurate simulation of the stress–strain behavior depends on both matrix nonlinearity and fiber–matrix debonding. Also, irrespective of the FOD (details of which have been described in section 3.2 and equation [2.2]), the primary source of nonlinearity during the early stage of testing (low applied loads) is fiber matrix debonding. The effect of matrix nonlinearity becomes progressively more dominant.

4.4. MSNC approach

Fatigue properties of SFRC are known to depend on the orientation and length distribution of the fibers (section 3.2), but the exact relationship is unknown. This is because unlike stiffness, which is well understood, the understanding of the fatigue behavior of composites is still rather limited. For a component-level simulation, each point in the FE model has a known FOD (and sometimes fiber length distribution [FLD] as well). RVEs, which represent the local FOD, can be created. Each point in the FE mesh can therefore be imagined as a different material whose stiffness and fatigue properties need to be estimated. As described in sections 4.2 and 4.3, the MT formulation provides good estimates of the effective stiffness as well as the stresses in individual fibers. Also, it has been shown that the MT formulation can adapted to model nonlinear stress–strain behavior during quasi-static loading. In this section, we build upon those models and also experimental observations (described in Chapter 3) to present a simple model for predicting the local SN curves as a function of the FOD.

Before we delve deeper into this topic, it is instructive to recall the concept of scaling of SN curves. Scaling of an SN curve refers to deriving the SN curve of RVE having variable FOD (or FLD) based on one reference SN curve. SN curves for RVE with different FODs are estimated by "scaling" up or down the reference SN curve based on calculated physical quantities such as endurance strength or UTS.

4.4.1. *Scaling of SN curves using the endurance limit*

A few publications have reported that the fatigue strength of SFRC is proportional to the UTS [BER 07a, DEM 12, ZAG 01]. Therefore, it seemed logical to attempt to scale the SN curve based on its UTS. However, such an approach has two major limitations. First, there is no physical reasoning as to why the fatigue strengths and the UTS must have same dependence on the FOD – particularly when there is experimental evidence that confirms that the failure mechanisms for damage propagation and final failure are different during tensile and fatigue loading [ARI 14a, HOR 96, ZAG 01]. The phenomena near final failure (catastrophic loss of load-carrying ability) during tensile failure are different from the phenomena during "slow" loss of load-carrying ability during cyclic loading. Second, this approach would require that the UTS for every FOD be known beforehand. Obviously, experimentally deriving UTS for every possible FOD is not feasible and to the best of our knowledge there are no reliable models for predicting the local UTS of SFRCs.

An alternative to scaling the SN curve based on the static strength could be to scale the SN curve based on the endurance strength (also sometimes referred to as fatigue limit). Endurance strength is defined as a load value below which the fatigue damage will "never" happen, or will happen with a very low probability. In the absence of true endurance strength for composites, it is common to consider the strength at 10^6 cycles. We first attempt to scale the SN curves based on the endurance strength which in turn is based on the assumption that onset of damage corresponds to the endurance strength of SFRC. This is similar to the argument made by [TAL 12] for different types of composite (see also Chapter 2, section 2.1). As described in section 4.3.1, unlike for tensile strength, calculating the onset of debonding (which is known to be the first damage event) is easy.

For 50% glass fiber reinforced PBT, SFRC coupons machined from plates (material and fatigue data described in section section 3.2), the stress levels at the onset of debonding were calculated for coupons machined in different directions with respect to the matrix flow direction. The predicted onset of debonding stress was based on the modified Coulomb's criteria (see equation [4.1]). It was seen that the stresses at which onset of debonding (either type A or type B depending on the primary fiber orientation with respect to loading direction) occurs are significantly lower than the expected endurance strength (Figure 4.10). It must be kept in mind that these calculations are based on the average stresses extracted using mean field homogenization techniques. As such the stress concentrations due to the presence of fiber in immediate neighborhood are not taken into account. Consequently, the predicted values of the debonding strength are actually upper bound values. It is expected that debonding actually starts at a much lower level of stress. Besides, the fatigue strengths of SFRC are known to drop as the angle between the fiber axis and loading direction changes. However, the stress at which onset of debonding occurred follows no such trend.

It can therefore be concluded that there is no correlation between the stress at which onset of fiber–matrix debonding occurs and the endurance strength of the SFRC. This is consistent with the conclusions made in Chapter 2 that the stress at which onset of damage occurs can be less than the endurance strength of long fiber composites.

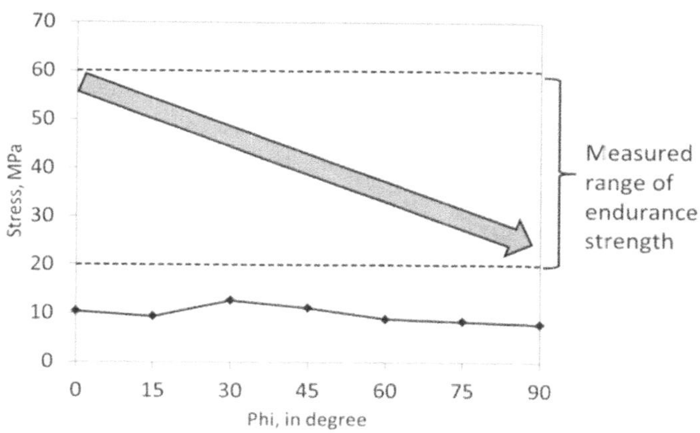

Figure 4.10. *Variation of the stress at which onset of debonding occurs for 50% wt. GF reinforced PBT for coupons machined in different angles (phi) with respect to the matrix flow direction. Dashed lines indicate the measured range of stress to failure after 10^6 cycles. Arrow indicates the expected trend for stress to failure after 10^6 cycles for coupons machined in different directions with respect to the flow of the matrix*

Fatigue failure of composites generally occurs when cracks are developed enough to cause final failure. Under constant amplitude cyclic loading, single fiber-matrix debond cannot propagate sufficiently to cause failure of SFRC. The extent of damage in SFRC must be sufficiently big to propagate and cause final failure. Thus, it was clear that the scaling approach based on the endurance strength or UTS is not feasible for predictions of the local SN curve based on the FOD.

The MSNC approach presented in section 4.4.2 is a solution for the above problems and can be used to determine the SN curves in a quick and reliable manner.

4.4.2. *MSNC approach*

The MSNC approach is based on the assumption that for SFRCs with different FOD and FLD (but same constituent), the extent of damage needed to propagate enough to cause final failure after a certain number of cycles must be the same. In other words, it is assumed that the damage propagation for two SFRC composites with the same constituents (fiber and the matrix) but different FODs will be the same, provided the number of cycles to failure is the same. The stress leading to failure after the number of cycles of loading is different for different FODs and is estimated by the proposed scheme. The extent of damage is quantified as the loss of secant stiffness in the applied loading direction. So, it is assumed that the loss of stiffness during cyclic loading in SFRC is independent of the FOD if the number of cycles to failure is the same.

Mathematically,

$$\frac{dE}{dN} = \text{const}(\text{FOD}) \qquad [4.3]$$

where E is the Young's modulus and N is the number of cycles.

It has been previously shown in section 3.2 that the loss of stiffness during cyclic loading is independent of the FOD for similar number of cycles. Using statistics, it has been demonstrated that the loss of stiffness curves can be assumed to be independent of the FOD. This is also the basic assumption of the MSNC approach.

A schematic representation of the proposed scheme is shown below in Figure 4.11.

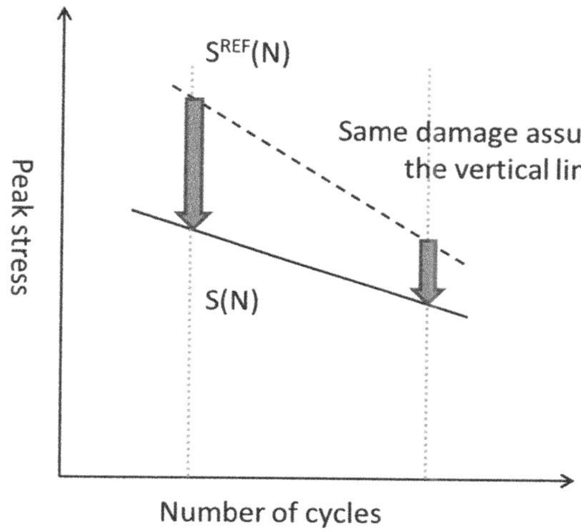

Figure 4.11. *A schematic representation of the MSNC approach [JAI 16b]*

Each data point in the SN curve gives information about the number of cycles to failure (x-axis) and the peak stress leading to failure after a corresponding number of cycles (y-axis). Based on the reference SN curve, the values of the damage parameter are calculated for the stress leading to failure after a certain number of cycles. It is then assumed that RVE having different FOD may have different stress to failure for the same number of cycles to failure, but where the damage parameter values remain the same. Thus, for different RVEs, for a given number of cycles to failure, the stress leading to failure is estimated by calculating the stress levels required to attain the same value of the damage parameter as was calculated from the input SN curve. Using this principle, the different points in the SN curves of different RVEs can be estimated and an SN curve can be generated by curve fitting.

A second interpretation of the MSNC could be as follows. During practical applications, often there are no exact methods to estimate the strength of a composite. In such a scenario, it is common to define the failure of a material when a certain predefined percentage of stiffness is lost. The proposed MSNC equates the loss of stiffness to generate the SN curve for the target RVE. Therefore, the MSNC can also be interpreted along the lines of scaling the SN curve based on the UTS. However, unlike scaling on the basis of UTS, the MSNC approach is based on equating lower values of stresses (corresponding to certain value of loss of stiffness).

Thus, one can expect the error in the MSNC predictions to be lower than the error during the scaling on the basis of UTS. The uncertainty in the modeling of the final tensile failure has been removed. Instead, the loss of stiffness due to the onset and propagation of the fiber–matrix debonding and the matrix nonlinearity is modeled. These quantities are easier to model and consequently there is greater confidence in the models for the loss of stiffness due to the onset and propagation of the fiber–matrix debonding.

A step-by-step description of the algorithm is given in the following and also in Figure 4.12:

– Step 1: Number of cycles to failure and corresponding peak stress, S1 value is extracted from the reference SN curve.

– Step 2: The Young's modulus of the test coupon, E_0, is calculated by the MT formulation.

– Step 3: Uniaxial loading is simulated till the peak stress, S1 by the MT formulation and the fiber–matrix debonding and matrix nonlinearity algorithms presented in section 4.3.

– Step 4: The Young's modulus of the reference test coupon after applied load, E_1, is calculated.

– Step 5: First cycle damage parameter, *d*, is calculated based on the following relation:

$$d = 1 - \frac{E_1}{E_0} \quad [4.4]$$

– Step 6: The initial Young's modulus of the target RVE (RVE whose SN curve is to be determined) is calculated.

– Step 7: Small load increments are applied to the second RVE. The stiffness and the damage parameter are calculated at the end of every time step.

– Step 8: Step 7 is repeated until the value of damage parameter is equal to *d*. The stress at which damage parameter is reached, S2 is the peak stress to failure for the chosen *N* number of cycles for the target RVE.

– Step 9: Steps 1–8 is repeated for at least three different numbers of cycles to failure and an SN curve is generated by using the fitting the calculated data points.

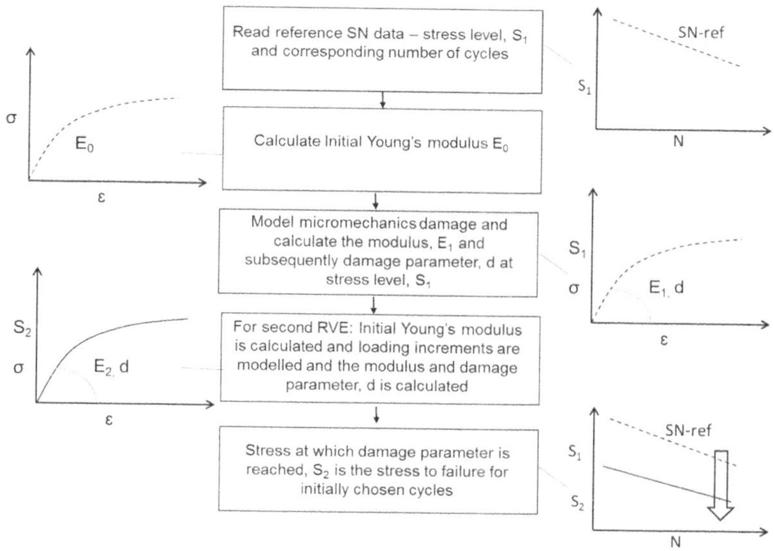

Figure 4.12. *A step by step schematic representation of the MSNC algorithm to generate local SN curves with one reference SN curve as input [JAI 16a]*

The input for the above algorithm is one reference SN curve, elastic properties of the fiber and the matrix, and the stress strain curve of the matrix. Of these required input, the stress strain curve of the neat resin is often the hardest to obtain. This is because SFRCs are made by injection molding precompounded pellets, which have both glass fiber and matrix in them. Thus, the stress–strain curve of the neat resin cannot be experimentally derived. Also, it is rare that the material supplier shares the exact configuration of the matrix. As shown in Figure 3.1, SFRCs made with the same thermoplastic but different polymer grade can have vastly different fatigue properties even though the fiber content and orientation is similar. Therefore, wrong stress strain data of the matrix can potentially lead to wrong results.

This difficulty of the MSNC approach can be circumvented by recalling two important conclusions, which were derived earlier in this book:

– the slope of the SN curve for an SFRC is independent of the FOD (see section 3.1.1 and Table 3.1);

– the predominant damage mode during low loads (applicable to high cycle fatigue) is fiber–matrix debonding (Figure 4.9).

Based on this, a second possibility is to ignore the matrix nonlinearity. Thus, the MSNC algorithm is based on the damage parameter, which is dependent on the fiber–matrix debonding only. The MSNC algorithm can then be implemented for only one point in the high cycle region. Since the applied load is low and the damage in the matrix is also low, ignoring matrix nonlinearity will not lead to a significant rise in error. After scaling one point on the basis of the MSNC curve, the rest of the SN curve can be generated assuming that the slope of the SN curve is independent of the FOD. This MSNC formulation is named "same-slope MSNC" or in short, "ss-MSNC". Since matrix nonlinearity is no longer modeled, the input for this formulation is one SN curve and the yield stress value of the pure matrix. For this approach, it is no longer necessary to have the stress–strain curve of the pure matrix. A schematic representation of the ss-MSNC scheme is shown in Figure 4.13.

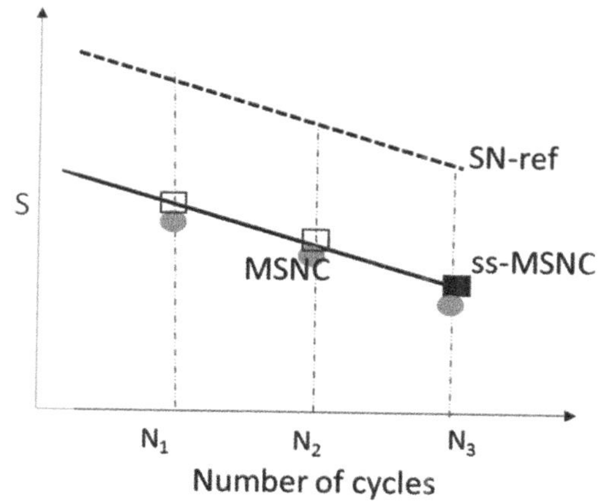

Figure 4.13. *Schematic representation of the ss-MSNC approach and comparison with the MSNC approach [JAI 16]*

4.4.3. *Validation*

4.4.3.1. *Sensitivity analysis*

The first step before the validation of a new model is to make sure that the number of fibers in the RVE is enough to make the calculations reliable. In other words, we must first make sure that the predictions by the proposed model will not change too much if the number of fibers is increased.

An SFRC with uniform random three-dimensional (3D) fiber orientation is expected to display quasi-isotropic behavior (same stiffness and fatigue properties at different directions). A certain number fibers are deemed to be high enough if both the stiffness and the fatigue strength predictions in all the directions of the isotropic SFRC are within a certain range of each other. Mathematically, the size of the RVE is said to be sufficient if:

$$\frac{\text{maximum\_predicted\_value}}{\text{minimum\_predicted\_value}} \leq 1.05 \qquad [4.5]$$

The ss-MSNC approach is implemented using different RVEs with varying numbers of fibers. The input for the calculations is taken to be the SN curve of the 0° coupon. SN curves are calculated for 81 different directions and the minimum and the maximum predicted values of the endurance strength are compared. The variation of the value of $\frac{\text{maximum\_predicted\_strength}}{\text{minimum\_predicted\_strength}}$ as a function of the chosen RVE size is shown in Figure 4.14.

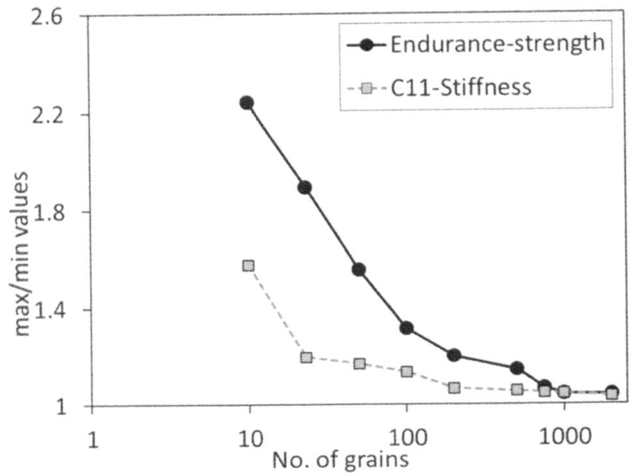

Figure 4.14. *Ratio of predicted maximum strength and minimum strength and stiffness C11 in different directions for a quasi-isotropic SFRC material made of 35% volume fraction GF reinforced with polybutylene terephthalate (PBT). The strength is calculated by the ss-MSNC approach while the stiffness is calculated by the MT formulation. Data from [JAI 16b]*

It is seen that there is a strong dependency of the stiffness and the MSNC results on the size of the RVE considered. The maximum strength was seen to be two to three times higher than the lowest strength, if an RVE with an insufficient number of fibers is considered. When 2,265 fibers are used, the scatter in the maximum and minimum values dropped to 1.03 and 1.04 for the stiffness and the endurance strength, respectively. It is thus confirmed that at least 1,000 fibers are needed to reliably use the MSNC approach. All the calculations henceforth in this book are performed with 1,000 fibers.

4.4.3.2. *Validation*

Using the data presented in Figure 3.6, the MSNC as well as the ss-MSNC approach will be validated. For both schemes, one SN curve at a time is taken as the reference SN curve and the other SN curves are predicted and compared with the experimental results in Figure 4.15. The validation is repeated for every SN curve reported. The MSNC approach is performed at three different numbers of cycles to failures: 10^4, 10^5, and 10^6 cycles. For the ss-MSNC approach, a stress and damage parameter at 10^6 cycles is considered and using same slope assumption the SN curves are generated. For all the calculations, an RVE size of 2,000 is considered. The RVE is created using the methods described by Onat and Leckie [ONA 88].

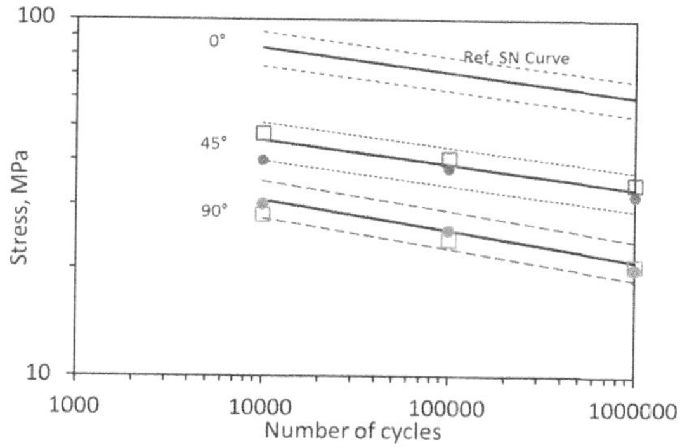

a)

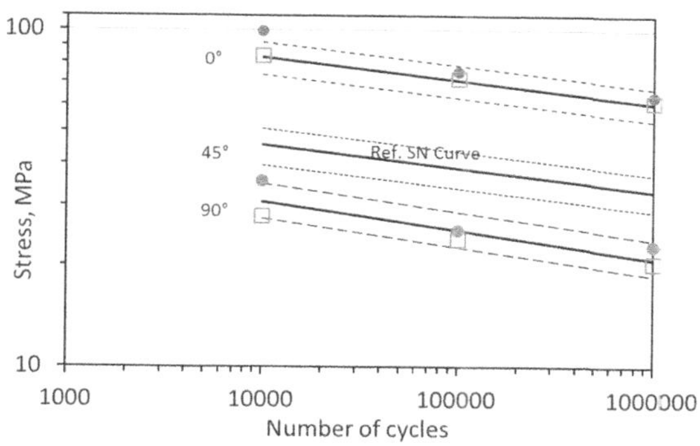

b)

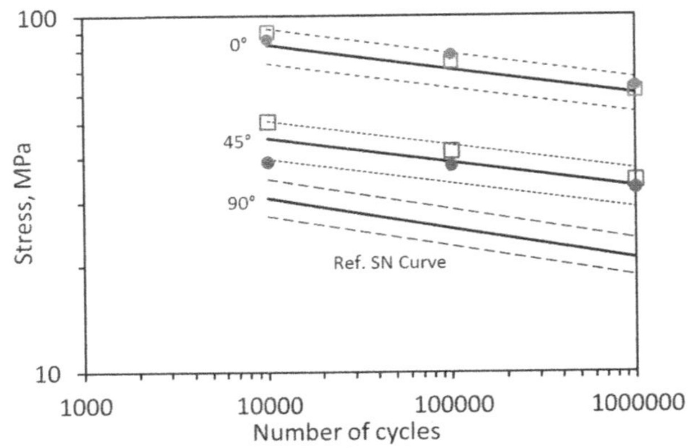

c)

Figure 4.15. *Validation of the MSNC and ss-MSNC approaches to generation of the local SN curves. In each simulation, one of the three SN curves is taken as input and the other two SN curves are used for validation of the proposed scheme. Hollow markers indicate results by ss-MSNC, while solid markers indicate MSNC results: a) 0° SN curve is reference, b) 45° SN curve is reference and c) 90° SN curve is reference. Data from [JAI 16a]*

The results of the validation are shown in Figure 4.15. It is seen that both the MSNC and ss-MSNC approach give good predictions of the SN curve, with the predicted SN curve within 90% confidence limit. It is also seen that accuracy of the proposed scheme did not depend on the FOD of the reference RVE.

Jain *et al.* [JAI 16a] validated both the MSNC and ss-MSNC approach for three sets of data and analyzed the error of the scheme both as a function of the FOD of the reference RVE as well as the number of cycles taken as input. After an ANOVA (Analysis of Variance) analysis of the errors, they concluded that the accuracy of both the MSNC as well as the ss-MSNC is independent of the FOD of the reference RVE. Also, extensions of the scheme to accommodate the skin–core variation of the FOD through the thickness were presented. This is a major argument in favor of the proposed approach as this reduces significantly the requirements for collecting test data. Test-based interpolation requires test coupons be somehow made with coupons having extreme FOD, which is not always possible.

Also, it was shown statistically that the ss-MSNC approach gave comparable results as compared to the MSNC approach. In fact, the accuracy of the ss-MSNC scheme at 10^5 cycles was found to be higher than the MSNC approach even though the input is less.

4.4.4. Comparison of the MSNC scheme with other schemes

In this section, the MSNC scheme is compared with two commonly used schemes. First, the scheme is compared with UTS-based scaling, followed by comparison with test-based interpolation.

4.4.4.1. *Comparison with UTS-based scaling*

The UTS-based scaling is compared to the proposed MSNC approach. UTS-based scaling is based on the assumption of constant proportionality between the static and fatigue strengths for different SFRC. Such an approach requires one SN curve as input and the UTS for every RVE, the UTS can either be experimentally derived for every FOD or modeled. The ratio of the UTS and the stress to failure for a certain number of cycles to failure is calculated from the reference SN curve and its UTS. The corresponding point in the SN curve for the target RVE is then calculated by multiplying the UTS for that RVE with the above calculated ratio. This process can be repeated for as many points as is deemed necessary to generate the SN curve.

Similar to the validation of the MSNC approach (presented in section 4.4.3), each of the three SN curves is taken as the reference SN curve and the other two SN curves are predicted. UTS-based scaling is tried for three number of cycles to failure, 10^4, 10^5 and 10^6 cycles (which is the same as the number of cycles chosen for the MSNC approach as well). The UTS for the 0°, 45°, and 90° coupons was taken to be 125, 59 and 47 MPa, respectively [JAI 16b].

The results of the UTS-based scaling are presented in Figure 4.16. The results of the MSNC approach are superimposed in the same figure for comparison. It is seen that the assumption of same proportionality between the static and fatigue strength is not reasonable. The predictions of the SN curve by this approach is seen to be sometimes even out of the 90% confidence intervals.

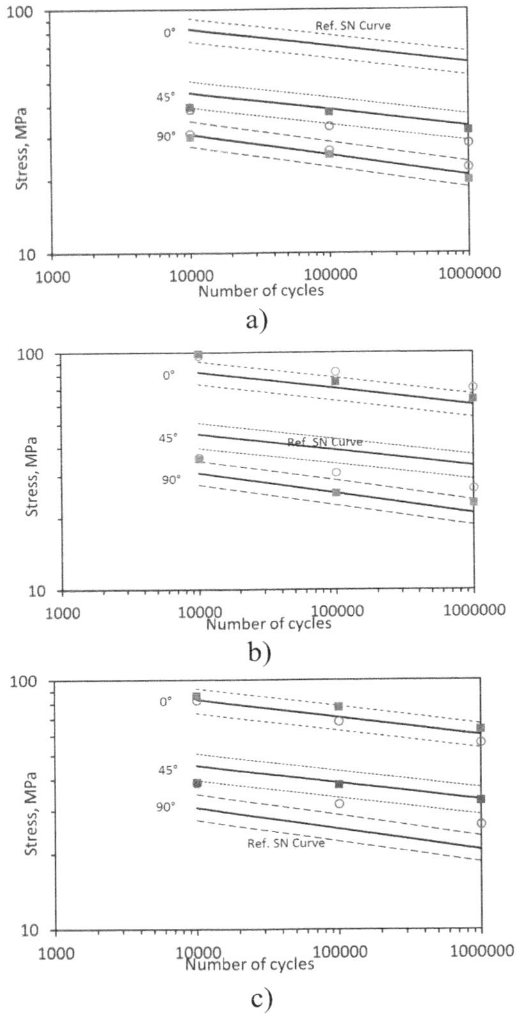

Figure 4.16. *Predictions of the UTS-based scaling (hollow round marker) and the MSNC approach (square marker) with one of the curves as the reference SN curve and the other two are simulated: a) SN curve of 0° coupon is taken as the reference SN curve; b) SN curve of 45° coupon is taken as the reference SN curve; c) SN curve of 90° coupon is taken as the reference SN curve. [JAI 16b]*

A closer look at the results reveals that the proportionality between the fatigue strengths and the static strengths is similar for the 0 and 90° coupons. This is confirmed by observing better predictions of the 90° SN curve when the 0° SN curve is taken as reference (and vice versa). However, the predicted SN curve for 45° coupon (and also when 45° SN curve is the reference SN curve) is outside of the 90% confidence interval limit. In contrast, the predictions of the MSNC approach are significantly better and more consistent for the three simulations considered. Therefore, it can be concluded that the MSNC approach is more accurate than UTS-based scaling.

4.4.4.2. *Comparison with test-based interpolation*

Next, the proposed scheme is compared with test-based interpolation methods. A test-based interpolation assumes a linear relationship between the SN curve and orientation of SFRC. Typically, SN curves are generated for SFRC with extreme orientation and the SN curves for intermediate orientations are estimated by interpolation. Contrary to the test-based interpolation, the MSNC approach predicted a nonlinear relationship between the orientation of the coupon and the fatigue property (Figure 4.15).

A test-based interpolation can be tried by the available data using the 0° and 90° SN curves as reference SN curves and the 45° SN curve is to be estimated by interpolation. Figure 4.17 compares the results of the test-based interpolation and the MSNC approach proposed in this book.

It is seen that a test-based interpolation overpredicts the fatigue strength of the 45° coupon. The three predicted points lie outside of the 90% confidence interval limits.

Apart from higher accuracy, the MSNC approach also provides significant benefits due to the easier testing required. Traditional interpolation-based approaches require several SN curves as input (three, if one assumes all fibers are oriented in the same plane and have the same length, and larger number of SN curves for a 3D FOD with varying lengths).

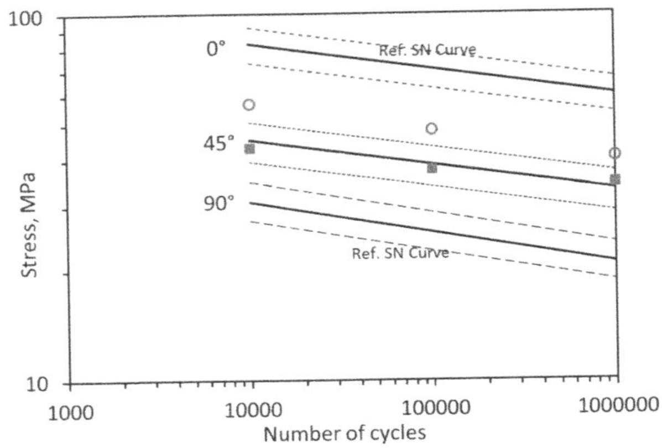

Figure 4.17. *Predictions of the test-based interpolation and the MSNC approach. For the test-based interpolation, the 0 and 90° SN curves are the input, while for the MSNC approach the 0° SN curve is the reference SN curve [JAI 16b]*

Three SN curves are needed if the interpolation of the SN curves are to be performed assuming that the fibers are orientated in the same plane. This is because a 2D fiber orientation can be represented by two terms of the second-order orientation tensor, i.e. a_{11} and a_{12}.

$$a_{2D} = \begin{bmatrix} a_{11} & a_{12} & 0 \\ a_{12} & 1 - a_{11} & 0 \\ 0 & 0 & 0 \end{bmatrix} \qquad [4.6]$$

For an interpolation of a quantity that depends on two variables, at least three points are needed as input. By this same reasoning, a 3D orientation of fiber can be represented by five terms of the orientation tensor and therefore six SN curves will be required for a proper interpolation.

Fatigue tests need to be performed on coupons with extreme FOD. Preparation of the coupons with extreme FOD is difficult and expensive. In contrast, the MSNC approach needs only one SN curve as input with no requirement on the FOD.

Interpolation-based approaches can take into account only one variable of the volume fraction, FLD or FOD, at a time. Additional tests on coupons with different FLDs and volume fractions must be performed if these factors are to be taken into account. The MSNC approach can be used even if the FLD and the FOD of the input SN curve is different from the RVE whose SN curve is desired.

4.5. Component-level simulations

There are very few reported attempts at component-level simulations for SFRC. The reason is that a complete fatigue simulation of SFRC will require a formulation to take into account all the factors listed in section 3.2. However, a single model to take into account every factor is yet to be formulated. Also, usually there is a complex geometry of the component leading to very complex FOD and the reliability of manufacturing simulation on such components is not very high [WHI 00].

Bernasconi *et al.* [BER 10b] performed simulations on a clutch pedal made of SFRC using experimental tests and simple FE-based analysis. Sonsino *et al.* [SON 08] formulated a design method that was based on the FE calculation of the maximum local stress and a fatigue analysis only on the highly stressed regions. This method was used to simulate the fatigue behavior of highly loaded engine compartments. Such an approach is suitable for metals but not for composites, since for SFRC the local lifetime to failure depends not only on the applied load but also the strengths in different directions. For an isotropic material, it is normally considered acceptable to look only at the highly stressed regions, but for composites such an approach will not work since regions of the SFRC component with low stress could also have low local strength (depending on the local FOD).

Hartmann *et al.* [HAR 11] performed fatigue simulations of a component and were able to correlate the fatigue tests on flat coupons with different applied R-ratio with the fatigue behavior of the component and showed good correlation with the experiments.

Malo *et al.* [MAL 13] performed simulation of long fiber SMCs using the pseudo-grain discretization method and failure criteria of

Liu and Mahadevan [LIU 07] and were able to predict the correctly critical points of the component (Figure 3.10) but the number of cycles to failure was not predicted.

Guster *et al.* [GUS 11] developed a method for fatigue life assessment of SFRC. They simulated an injection molded ring spanner made of short glass fiber reinforced polyamide-6T/6I. Fiber orientation and distribution was taken into account by introducing correction factors. They reported a good agreement with the maximum damage areas but the lifetime estimation was off by about a factor of 3.

Vervoort [VER 13] estimated the fatigue life of a component on the bottom of the spare wheel compartment of a car made up of short fiber reinforced polypropylene. They performed two sets of simulations; first using isotropic properties and the other with orthotropic properties based on the second-order orientation tensor predicted by manufacturing simulation. There was significant difference in the predicted critical locations by the two simulations, thus confirming the need for accounting for the local FOD and the different static and fatigue properties at every point. Vervoort generated the SN curves at different points based on three experimental SN curves. He was able to correctly predict the critical area of the specimen but the number of cycles to failure was not predicted.

We now proceed to show an example of component-level simulation performed using the MSNC approach. The ss-MSNC approach was implemented in software LMS Virtual.Lab Durability [LMS 15] using special links with multiscale platform Converse [CON 14].

Multiaxial loads are accounted for using the so-called "characteristic plane approach". As a part of this approach, 81 SN curves are generated for each element. These 81 SN curves represent the SN curves for uniaxial load in 81 different directions. The lifetime of an element is then determined by rotating the stress tensor to those 81 directions and then determining the lowest life. The characteristic plane of the material was defined as the plane in which the material is the weakest and was found to be a function of the material properties as well as the applied load. The loss of stiffness during cyclic loading is accounted for by using the proprietary hysteresis operator patented by Siemens' industry software [BRU 14]. Apart from the fatigue solver LMS Virtual.Lab Durability, manufacturing simulation software Sigmasoft [SIG 14] and FE software from SAMTECH were used for the fatigue simulation.

To validate the proposed model, a representative industrial component, nicknamed Pinocchio, is considered, special test rigs are designed and the component is subject to fatigue loading (Figure 4.18).

a)

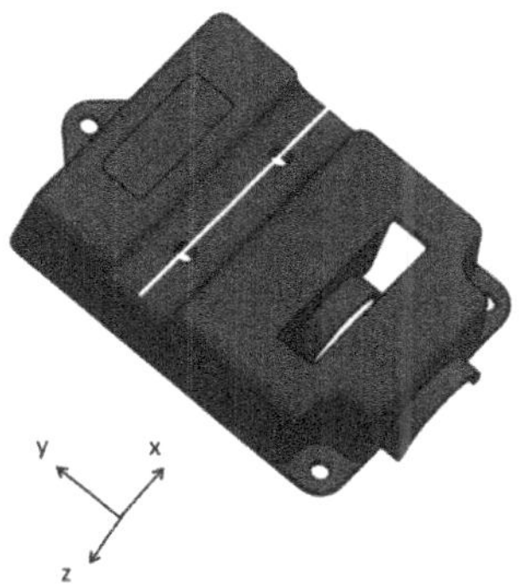

b)

Figure 4.18. *Test set up and loading conditions for fatigue testing of "Pinocchio": a) overview figure where 1 is the rig, 2 is the component, 3 are the screws and 4 is the load actuator with indenter; b) isometric view of the component "Pinocchio", the load actuator hits the component on the white line*

A horizontal Schenck machine equipped with a load cell of 10 kN was used for the fatigue tests. The tests were load controlled and a sinusoidal load function with constant amplitude was applied. The applied loads had an R-ratio of 0.1. The criterion to end the fatigue tests was complete specimen separation into two or more parts, penetration of the load actuator into the specimen or no breakage until 10^6 cycles.

Comparison of the experimental and simulations show that the developed methodology for fatigue simulation presents a reasonably good match at high cycle fatigue. Also, the proposed method was able to detect the correct area for failure.

At low cycle fatigue, it is not enough to simply model failure by considering loss of stiffness; some consideration must also be given to the fracture mechanisms in the component (see Figure 4.19).

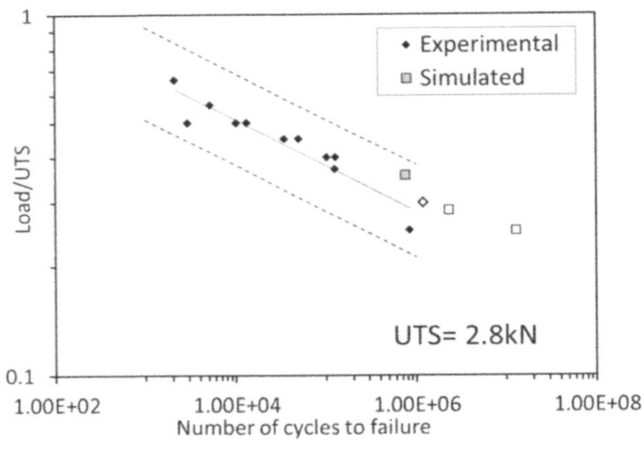

Figure 4.19. *Simulated SN curve and comparison with the experimental results for component "Pinocchio"*

4.6. Conclusions and future outlook

A method for simulation of SFRC fatigue was developed. Different aspects of SFRC simulation such as homogenization, damage modeling and fatigue modeling were studied, and prevalent practices were critically examined. A hybrid multiscale method to predict the fatigue behavior of SFRC has been developed in this book; this approach is called the "Master SN curve approach". The proposed method is based on material tests, descriptions of the composite internal geometry (via manufacturing simulation) and multiscale mechanics.

A building block to development of the fatigue model was presented. It was first shown that full MT formulation has superior predictive capabilities to the more prevalent PGMT. Next elaborate models for the damage in SFRC materials were presented. Based on the above two steps, a MSNC approach to fatigue simulation has been presented.

It has been shown that the MSNC approach is a reliable method for predicting the local SN curves as a function of the FOD. Since, this scheme is based on the MT formulation, this is not too computationally expensive and can be used to generate large number of SN curves. However, it must be noted that the SN curve generated by this scheme has the same applied load characteristics as that of the reference SN curve. The MSNC approach has been designed to work when the reference SN curve is generated using test data for uniaxial tension–tension loading with R-ratio close to 0. The characteristics of the SN curve calculated using the MSNC approach is also the same. However, during component-level loading each element will be subject to multiaxial loading, uniaxial loading is rather exceptional during real life applications. Simple models to account for multiaxial fatigue have been developed. These models are rather limited in scope and future attempts must be devoted to extending the present methods to include different factors like multiaxial loading, different load ratios and loading at different temperatures.

4.7. Bibliography

[ABD 15a] ABDIN Y., Mechanical properties and fatigue of random steel fibre reinforced composites, Thesis, University of Leuven, 2015.

[ABD 15b] ABDIN Y., LOMOV S.V., JAIN A. *et al.*, "Mean-field based fatigue damage modeling of composites reinforced with short straight and wavy fibers", *20th International Conference on Composite Materials (ICCM-20)*, Copenhagen, 2015.

[ADV 87] ADVANI S.G., TUCKER C.L., "The use of tensors to describe and predict fiber orientation in short fiber composites", *Journal of Rheology*, vol. 31, no. 8, pp. 751–784, 1987.

[ARI 14a] ARIF M.F., SAINTIER N., MERAGHNI F. *et al.*, "Multiscale fatigue damage characterization in short glass fiber reinforced polyamide-66", *Composites Part B - Engineering*, vol. 61, pp. 55–65, 2014.

[ARI 14b] ARIF M.F., MERAGHNI F., CHEMISKY Y. *et al.*, "In-situ damage mechanisms investigation of pa66/gf30 composite: effect of relative humidity", *Composites Part B - Engineering*, vol. 58, pp. 487–495, 2014.

[AUT 16] AUTODESK, "Autodesk Moldflow", available at www.autodesk.com/moldflow, 2016.

[BER 07a] BERNASCONI A., DAVOLI P., BASILE A. *et al.*, "Effect of fibre orientation on the fatigue behaviour of a short glass fibre reinforced polyamide-6", *International Journal of Fatigue*, vol. 29, no. 2, pp. 199–208, 2007.

[BER 07b] BERNASCONI A., DAVOLI P., ROSSIN D. *et al.*, "Effect of reprocessing on the fatigue strength of a fibre glass reinforced polyamide", *Composites Part A-Applied Science and Manufacturing*, vol. 38, no. 3, pp. 710–718, 2007.

[BER 10a] BERNASCONI A., COSMI F., ZAPPA E., "Combined effect of notches and fibre orientation on fatigue behaviour of short fibre reinforced polyamide", *Strain*, vol. 46, no. 5, pp. 435–445, 2010.

[BER 10b] BERNASCONI A., DAVOLI P., ARMANNI C., "Fatigue strength of a clutch pedal made of reprocessed short glass fibre reinforced polyamide", *International Journal of Fatigue*, vol. 32, pp. 100–107, 2010.

[BRU 14] BRUYNEEL M., DONDERS S., HACK M. *et al.*, Calculating fatigue and fatigue failure of structures, Patent US 20140336954 A1, 2014.

[CAM 90] CAMACHO C.W., TUCKER C.L., YALVAC S. *et al.*, "Stiffness and thermal-expansion predictions for hybrid short fiber composites", *Polymer Composites*, vol. 11, no. 4, pp. 229–439, 1990.

[CIN 95] CINTRA J.S., TUCKER C.L., "Orthotropic closure approximations for flow-induced fiber orientation", *Journal of Rheology*, vol. 39, no. 6, pp. 1095–1122, 1995.

[CON 14] CONVERSE, "CONVERSE: process structure interaction for injection molded plastic parts", available at http://partengineering.com/, 2014.

[COS 11] COSMI F., BERNASCONI A., SODINI N., "Phase contrast micro-tomography and morphological analysis of a short carbon fibre reinforced polyamide", *Composites Science and Technology*, vol. 71, no. 1, pp. 23–30, 2011.

[COS 13] COSMI F., BERNASCONI A., "Micro-CT investigation on fatigue damage evolution in short fibre reinforced polymers", *Composites Science and Technology*, vol. 79, pp. 70–76, 2013.

[DEM 10] DE MONTE M., MOOSBRUGGER E., QUARESIMIN M., "Influence of temperature and thickness on the off-axis behaviour of short glass fibre reinforced polyamide 6.6-quasi-static loading", *Composites Part A-Applied Science and Manufacturing*, vol. 41, no. 7, pp. 859–871, 2010.

[DEM 12] DE MONTE M., MOOSBRUGGER E., QUARESIMIN M., "Influence of temperature and thickness on the off-axis behaviour of short glass fibre reinforced polyamide 6.6 cyclic loading", *Composites Part A: Applied Science and Manufacturing*, vol. 41, no. 10, pp. 1368–1379, 2012.

[DEG 01] DEGRIECK J., VAN PAEPEGEM W., "Fatigue damage modeling of fibre-reinforced composite materials: review", *Applied Mechanics Reviews*, vol. 54, no. 4, pp. 279–300, 2001.

[DIG 12] DIGIMAT, "DIGIMAT: The multi-scale modelling platform", available at: www.e-Xstream.com, 2012.

[DOG 03] DOGHRI I., OUAAR A., "Homogenization of two-phase elasto-plastic composite materials and structures – study of tangent operators, cyclic plasticity and numerical algorithms", *International Journal of Solids and Structures*, vol. 40, no. 7, pp. 1681–1712, 2003.

[DOG 05] DOGHRI I., TINEL L., "Micromechanical modeling and computation of elasto-plastic materials reinforced with distributed-orientation fibers", *International Journal of Plasticity*, vol. 21, no. 10, pp. 1919–1940, 2005.

[EBE 01] EBERHARDT C., CLARKE A., VINCENT M. *et al.*, "Fibre-orientation measurements in short-glass-fibre composites-ii: a quantitative error estimate of the 2D image analysis technique", *Composites Science and Technology*, vol. 61, no. 13, pp. 1961–1974, 2001.

[ERH 99] ERHARD G., *Designing with Polymers*, Carl Hanser Publishing, Munich–Vienna, 1999.

[ESH 57] ESHELBY J.D., "The determination of the elastic field of an ellipsoidal inclusion, and related problems", *Proceedings of the Royal Society of London Series A - Mathematical and Physical Sciences*, vol. 241, no. 1226, pp. 376–396, 1957.

[FAR 04] FARA S., PAVAN A., "Fibre orientation effects on the fracture of short fibre polymer composites: on the existence of a critical fibre orientation on varying internal material variables", *Journal of Materials Science*, vol. 39, no. 11, pp. 3619–3628, 2004.

[GHO 12] GHOSSEIN E., LEVESQUE M., "A fully automated numerical tool for a comprehensive validation of homogenization models and its application to spherical particles reinforced composites", *International Journal of Solids and Structures*, vol. 49, nos. 11–12, pp. 1387–1398, 2012.

[GUP 90] GUPTA V.B., MITTAL R.K., GOEL M., "Energy absorbing mechanisms in short-glass-fiber-reinforced polypropylene", *Composites Science and Technology*, vol. 37, no. 4, pp. 353–369, 1990.

[GUS 97] GUSEV A., "Representative volume element size for elastic composites: a numerical study", *Journal of the Mechanics and Physics of Solids*, vol. 45, no. 9, pp. 1449–1459, 1997.

[GUS 11] GUSTER C., PINTER G., MOSENBACHER A. *et al.*, "Evaluation of a simulation process for fatigue life calculation of short fibre reinforced plastic components", *11th International Conference on the Mechanical Behavior of Materials (ICM11)*, Milan, Italy, 2011.

[HAL 78] HALPIN J.C., KARDOS J.L., "Strength of discontinuous reinforced composites .1. Fiber reinforced composites", *Polymer Engineering and Science*, vol. 18, no. 6, pp. 496–504, 1978.

[HAR 11] HARTMANN J., MOOSBRUGGER E., Buter A., "Variable amplitude loading with components made of short fiber reinforced polyamide 6.6", *Procedia Engineering*, vol. 10, pp. 2009–2015, 2011.

[HIL 65] HILL R., "A self-consistent mechanics of composite materials", *Journal of the Mechanics and Physics of Solids*, vol. 13, no. 4, pp. 213–222, 1965.

[HIN 95] HINE P.J., DAVIDSON N., DUCKETT R.A. *et al.*, "Measuring the fiber orientation and modeling the elastic properties of injection-molded long-glass-fiber-reinforced nylon", *Composites Science and Technology*, vol. 53, no. 2, pp. 125–131, 1995.

[HOR 96] HORST J.J., SPOORMAKER J.L., "Mechanisms of fatigue in short glass fiber reinforced polyamide 6", *Polymer Engineering and Science*, vol. 36, no. 22, pp. 2718–2726, 1996.

[HOR 97] HORST J.J., SPOORMAKER J.L., "Fatigue fracture mechanisms and fractography of short-glass fibre-reinforced polyamide 6", *Journal of Materials Science*, vol. 32, no. 14, pp. 3641–3651, 1997.

[HUY 01] HUYSMANS G., VERPOEST I., VAN HOUTTE P., "A damage model for knitted fabric composites", *Composites Part A - Applied Science and Manufacturing*, vol. 32, no. 10, pp. 1465–1475, 2001.

[JAC 04] JACK D.A., SMITH D.E., "Assessing the use of tensor closure methods with orientation distribution reconstruction functions", *Journal of Composite Materials*, vol. 38, no. 21, pp. 1851–1871, 2004.

[JAI 13] JAIN A., LOMOV S.V., ABDIN Y. *et al.*, "Pseudo-grain discretization and full Mori-Tanaka formulation for random heterogeneous media: predictive abilities for stresses in individual inclusions and the matrix", *Composites Science and Technology*, vol. 87, pp. 86–93, 2013.

[JAI 15a] JAIN A., ABDIN Y., VAN PAEPEGEM W. *et al.*, "Effective anisotropic stiffness of inclusions with debonded interface for Eshelby-based models", *Composite Structures*, vol. 131, pp. 692–706, 2015.

[JAI 15b] JAIN A., ABDIN Y., VAN PAEPEGEM W. *et al.*, "Non-symmetric stiffness tensor prediction by the Mori-Tanaka scheme – comments on the article 'effective anisotropic stiffness of inclusions with debonded interface for Eshelby-based models'", *Composite Structures*, vol. 134, pp. 1118–1119, 2015.

[JAI 16a] JAIN A., VAN PAEPEGEM W., VERPOEST I. *et al.*, "A feasibility study of the master SN curve approach for short fiber reinforced composites", *International Journal of Fatigue*, vol. 91, pp. 264–274, 2016.

[JAI 16b] JAIN A., VEAS J.M., STRAESSER S. *et al.*, "The master SN curve approach – a hybrid multiscale fatigue simulation of short fiber reinforced composites", *Composites Part A: Applied Science and Manufacturing*, vol. 91, no. 2, pp. 510–518, 2016.

[JAO 05] JAO JULES E., LOMOV S., VERPOEST I. *et al.*, "Prediction of non-linear behaviour of discontinuous long glass fibres polypropylene composites", *15th International Conference of Composite Materials (ICCM-15)*, Durban, 2005.

[JEG 13] JEGOU L., MARCO Y., LE SAUX V. *et al.*, "Fast prediction of the wohler curve from heat build-up measurements on short fiber reinforced plastic", *International Journal of Fatigue*, vol. 47, pp. 259–267, 2013.

[KAR 85] KARDOS J.L., "Critical issues in achieving desirable mechanical properties for short fiber composites", *Pure and Applied Chemistry*, vol. 57, no. 11, pp. 1651–1657, 1985.

[KAR 07] KARI S., BERGER H., RODRIGUEZ-RAMOS R. *et al.*, "Numerical evaluation of effective material properties of transversely randomly distributed unidirectional piezoelectric fiber composites", *Journal of Intelligent Material Systems and Structures*, vol. 18, no. 4, pp. 361–372, 2007.

[KLI 11] KLIMKEIT B., NADOT Y., CASTAGNET S. *et al.*, "Multiaxial fatigue life assessment for reinforced polymers", *International Journal of Fatigue*, vol. 33, no. 6, pp. 766–780, 2011.

[KLI 13] KLIMKEIT B., BERGAMO S., NADOT Y. *et al.*, "Multiaxial fatigue behavior of short-fiber reinforced thermoplastic roof bars", *International Conference on Multiaxial Fatigue & Fracture (ICMFF9)*, Parma, Italy, 2013.

[LAU 13] LAUNAY A., MAITOURNAM M.H., MARCO Y. *et al.*, "Multiaxial fatigue models for short glass fiber reinforced polyamide – part I: nonlinear anisotropic constitutive behavior for cyclic response", *International Journal of Fatigue*, vol. 47, pp. 382–389, 2013.

[LIE 99] LIELENS G., Micro-macro modeling of structured materials, Thesis, University of Leuven, 1999.

[LIU 07] LIU Y., MAHADEVAN S., "A unified multiaxial fatigue damage model for isotropic and anisotropic materials", *International Journal of Fatigue*, vol. 29, pp. 347–359, 2007.

[LMS 15] LMS, "LMS Virtual.Lab durability-durability analysis for optimal product performance", available at: http://www.plm.automation.siemens.com/en_us/products/lms/virtual-lab/durability/, 2015.

[MAL 13] MALO T., ADAM L., ASSAKER R. *et al.*, "Multi-scale modeling of high cycle fatigue of chopped and continuous fiber composites", *Society of Plastic Engineers (SPE) Automotive*, Michigan, USA, 2013.

[MEN 10] MENEGHETTI G., QUARESIMIN M., "Fatigue strength assessment of a short fiber composite based on the specific heat dissipation", *Composites Part B: Engineering*, vol. 42, no. 2, pp. 217–225, 2010.

[MER 95] MERAGHNI F., BENZEGGAGH M.L., "Micromechanical modeling of matrix degradation in randomly oriented discontinuous-fiber composites", *Composites Science and Technology*, vol. 55, no. 2, pp. 171–186, 1995.

[MIE 12] MIETTINEN A., HENDRIKS C.L.L., CHINGA-CARRASCO G. *et al.*, "A non-destructive X-ray microtomography approach for measuring fibre length in short-fibre composites", *Composites Science and Technology*, vol. 72, no. 15, pp. 1901–1908, 2012.

[MLE 99] MLEKUSCH B., "Fibre orientation in short-fibre-reinforced thermoplastics ii. quantitative measurements by image analysis", *Composites Science and Technology*, vol. 59, no. 4, pp. 547–560, 1999.

[MOR 73] MORI T., TANAKA K., "Average stress in matrix and average elastic energy of materials with misfitting inclusions", *Acta Metallurgica*, vol. 21, no. 5, pp. 571–574, 1973.

[MOS 14] MOSENBACHER A., BRUNBAUER J., PICHLER P.F. *et al.*, "Modelling and validation of fatigue life calculation method for short fiber reinforced injection molded parts", *16th European Conference of Composite Materials (ECCM-16)*, Seville, Spain, 2014.

[MOU 06] MOUHMID B., IMAD A., BENSEDDIQ N. *et al.*, "A study of the mechanical behaviour of a glass fibre reinforced polyamide 6,6: experimental investigation", *Polymer Testing*, vol. 25, no. 4, pp. 544–552, 2006.

[ONA 88] ONAT E.T., LECKIE F.A., "Representation of mechanical behavior in the presence of change internal structure", *Journal of Applied Mechanics*, vol. 55, no. 1, pp. 1–10, 1988.

[PAP 97] PAPADOPOULOS I.V., DAVOLI P., GORLA C. *et al.*, "A comparative study of multiaxial high-cycle fatigue criteria for metals", *International Journal of Fatigue*, vol. 19, no. 3, pp. 219–235, 1997.

[PIE 04] PIERARD O., FRIEBEL C., DOGHRI I., "Mean-field homogenization of multi-phase thermo-elastic composites: a general framework and its validation", *Composites Science and Technology*, vol. 64, no. 10–11, pp. 1587–1603, 2004.

[PIE 07] PIERARD O., GONZALEZ C., SEGURADO J. *et al.*, "Micromechanics of elasto-plastic materials reinforced with ellipsoidal inclusions", *International Journal of Solids and Structures*, vol. 44, no. 21, pp. 6945–6962, 2007.

[QUA 10] QUARESIMIN M., SUSMEL L., TALREJA R., "Fatigue behaviour and life assessment of composite laminates under multiaxial loadings", *International Journal of Fatigue*, vol. 32, no. 1, pp. 2–16, 2010.

[REI 91] REIFSNIDER K.L., GAO Z., "A micromechanics model for composites under fatigue loading", *International Journal of Fatigue*, vol. 13, no. 2, pp. 149–156, 1991.

[RIN 97] RINTOUL M.D., TORQUATO S., "Reconstruction of the structure of dispersions", *Journal of Colloid and Interface Science*, vol. 186, no. 2, pp. 467–476, 1997.

[SAT 91] SATO N., KURAUCHI T., SATO S., "Microfailure behavior of randomly dispersed short fiber reinforced thermoplastic composites obtained by direct SEM observation", *Journal of Materials Science*, vol. 26, no. 14, 3891–3898, 1991.

[SIG 14] SIGMA, "SIGMA Engineering GmbH", available at http://www.sigmasoft.de/, 2014.

[SON 08] SONSINO C.M., MOOSBRUGGER E., "Fatigue design of highly loaded short-glass-fibre reinforced polyamide parts in engine compartments", *International Journal of Fatigue*, vol. 30, pp. 1279–1288, 2008.

[SPA 14] SPANCKEN D., DECKER J., BECKER B. *et al.*, "Fatigue design for short fiber reinforced polymers", *16th European Conference of Composite Materials (ECCM-16)*, Seville, Spain, 2014.

[TAL 12] TALREJA R., SINGH C.V., *Damage and Failure of Composite Materials*, Cambridge University Press, 2012.

[TAN 88] TANDON G.P., WENG G.J., "A theory of particle-reinforced plasticity", *Journal of Applied Mechanics*, vol. 55, no.1, pp. 126–135, 1988.

[THO 02] THOMASON J.L., "Interfacial strength in thermoplastic composites - at last an industry friendly measurement method?", *Composites Part A-Applied Science and Manufacturing*, vol. 33, no. 10, pp. 1283–1288, 2002.

[TUC 99] TUCKER III, CHARLES L., ERWIN LIANG *et al.*, "Stiffness predictions for unidirectional short-fiber composites: review and evaluation", *Composites Science and Technology*, vol. 59, no. 5, pp. 655–671, 1999.

[VER 13] VERVOORT S., "Fatigue analysis of polymer components with short fiber reinforcements", *NAFEMS World Congress*, Salzburg, Austria, 2013.

[WHI 00] WHITESIDE B.R., COATES P.D., HINE P.J. *et al.*, "Glass fibre orientation within injection moulded automotive pedal: simulation and experimental studies", *Plastic Rubber Composites*, vol. 29, pp. 38–45, 2000.

[WYZ 04] WYZGOSKI M.G., KROHN J.A., NOVAK G.E., "Fatigue of fiber-reinforced injection molded plastics. I: stress-lifetime data", *Polymer Composites*, vol. 25, no. 5, 489–498, 2004.

[WYZ 05] WYZGOSKI M.G., NOVAK G.E., "Predicting fatigue s-n (stress-number of cycles to fail) behavior of reinforced plastics using fracture mechanics theory", *Journal of Materials Science*, vol. 40, no. 2, pp. 295–308, 2005.

[ZAG 01] ZAGO A., SPRINGER G.S., "Fatigue lives of short fiber reinforced thermoplastics parts", *Journal of Reinforced Plastics and Composites*, vol. 20, no. 7, pp. 606–620, 2001.

[ZHO 05] ZHOU Y., MALLICK P.K., "A non-linear damage model for the tensile behavior of an injection molded short e-glass fiber reinforced polyamide-6, 6", *Materials Science and Engineering: A*, vol. 393, no. 1, pp. 303–309, 2005.

[ZHO 13] ZHOU H., *Computer Modeling for Injection Molding: Simulation, Optimization, and Control*, Wiley, 2013.

Index

Q, S

T, W

Other titles from

ISTE

in

Materials Science

2016

TORRENTI Jean-Michel
Materials and Infrastructures 1

Materials and Infrastructures 2

RÉMOND Yves, AHZI Saïd, BANIASSADI Majid, GARMESTANI Hamid
Applied RVE Reconstruction and Homogenization of Heterogeneous Materials

2015

JOUD Jean-Charles, BARTHÉS-LABROUSSE Marie-Geneviève
Physical Chemistry and Acid-Base Properties of Surfaces

LALEVÉE Jacques, FOUASSIER Jean-Pierre
Dyes and Chromophores in Polymer Science

2014

HAMAIDE Thierry, DETERRE Rémi, FELLER Jean-François
Environmental Impact of Polymers

LAFLEUR Pierre, VERGNES Bruno
Polymer Extrusion

2013

BATHIAS Claude
Fatigue Limit in Metals

LEXCELLENT Christian
Shape-memory Alloys Handbook

2012

DAVIM J. Paulo
Wear of Advanced Materials

DELHAES Pierre
Carbon Science and Technology: From Energy to Materials

GOUDEAU Philippe, GUINEBRETIÈRE René
X-Rays and Materials

SAANOUNI Khemais
Damage Mechanics in Metal Forming: Advanced Modeling and Numerical Simulation

VANBÉSIEN Olivier
Artificial Materials

VERDU Jacques
Oxidative Ageing of Polymers

2011

BATHIAS Claude, PINEAU André
Fatigue of Materials and Structures: Application to Damage

BATHIAS Claude, PINEAU André
Fatigue of Materials and Structures: Application to Design

PRIESTER Louisette
Grain Boundaries and Crystalline Plasticity

VIGNES Alain
Extractive Metallurgy 1: Basic Thermodynamics and Kinetics
Extractive Metallurgy 2: Metallurgical Reaction Processes
Extractive Metallurgy 3: Processing Operations and Routes

2010

BATHIAS Claude, PINEAU André
Fatigue of Materials and Structures: Fundamentals

CHATEIGNER Daniel
Combined Analysis

CHEVALIER Yvon, VINH TUONG Jean
Mechanical Characterization of Materials and Wave Dispersion V1
Mechanical Characterization of Materials and Wave Dispersion V2

DELHAES Pierre
Carbon-based Solids and Materials

2009

ALVAREZ-ARMAS Iris, DEGALLAIX-MOREUIL Suzanne
Duplex Stainless Steels

DAVIM J. Paulo
Machining Composite Materials

GALERIE Alain
Vapor Surface Treatments

2008

BATTAGLIA Jean-Luc
Heat Transfer in Materials Forming Processes

BLONDEAU Régis
Metallurgy and Mechanics of Welding

FRANÇOIS Dominique
Structural Components

REYNE Maurice
Plastic Forming Processes

TAKADOUM Jamal
Materials and Surface Engineering in Tribology

2007

BOCH Philippe, NIEPCE Jean-Claude
Ceramic Materials

CRISTESCU Constantin
Materials with Rheological Properties

2006

LASSEN Tom, RÉCHO Naman
Fatigue Life Analyses of Welded Structures

Printed and bound by CPI Group (UK) Ltd, Croydon, CR0 4YY
16/09/2026
14963078-0001